HIGHWAY ENGINEERING

Edited by **Hamid Yaghoubi**

Highway Engineering
http://dx.doi.org/10.5772/intechopen.68229
Edited by Hamid Yaghoubi

Contributors

Carlo Polidori, Adewole Adesiyun, Liljana Cela, Antonio Avenoso, Christophe Nicodème, Kallistratos Dionelis, Ilker Erkan, Gunawan Handayani, Ozgur Baskan, Cenk Ozan, Martin Vlkovský, Mohammed Ali Berawi

Published by InTech

Janeza Trdine 9, 51000 Rijeka, Croatia

Publishing Process Manager Iva Simcic
Technical Editor SPi Global
Cover InTech Design team

First published December, 2017
Printed in Croatia
Legal deposit, Croatia: National and University Library in Zagreb

Additional hard copies can be obtained from orders@intechopen.com

Highway Engineering, Edited by Hamid Yaghoubi
p. cm.
Print ISBN 978-953-51-3669-9
Online ISBN 978-953-51-3670-5

For EU product safety concerns:
IN TECH d.o.o., Prolaz Marije Krucifikse Kozulić 3, 51000 Rijeka, Croatia,
info@intechopen.com or visit our website at intechopen.com.

Meet the editor

Dr. Hamid Yaghoubi is the director of Iran Maglev Technology (IMT), Iran. He became the Iran top researcher in 2010. In this regard, he was awarded by the Iranian President, Iranian Minister of Science, Research and Technology, and Iranian Minister of Information and Communications Technology. Dr. Yaghoubi became the 2011 and 2012 Outstanding Reviewer for the *Journal of Transportation Engineering (JTE)* and American Society of Civil Engineers (ASCE), USA. One of his journal papers became as the 2011 Top Download Paper for JTE. He received the ICCTP2011 award for the 11th International Conference of Chinese Transportation Professionals (ICCTP2011), ASCE. He is an assistant chief editor and editorial board member for some journals. He has been a reviewer for the majority of journals, books, and conferences. Dr. Yaghoubi has cooperated with hundreds of international conferences as chairman, keynote speaker, chair of session, publication chair, and member of committees, including scientific, organizing, steering, advisory, technical program, etc. He is also a member in several international committees.

Contents

Preface IX

Chapter 1 **Reserve Capacity Model for Optimizing Traffic Signal Timings with an Equity Constraint 1**
Ozgur Baskan and Cenk Ozan

Chapter 2 **Effects on the Design of Transport Systems of Pedestrian Dynamics 17**
İlker Erkan

Chapter 3 **Analysis of Highway Acceleration in Regard to Cargo Security 35**
Martin Vlkovský

Chapter 4 **Joint Road Safety Analysis in Open Roads and Tunnels 55**
Carlo Polidori, Adewole Adesiyun, Antonio Avenoso, Kallistratos Dionelis, Liljana Cela, Christophe Nicodème and Thierry Goger

Chapter 5 **Empowering Added Value in Highway Project: A Strategy to Improve the Feasibility 67**
Mohammed Ali Berawi

Chapter 6 **Nondestructive Tests in Highway Engineering 87**
Gunawan Handayani

Preface

Highway engineering is an engineering discipline branching from civil engineering that involves the planning, design, construction, operation, and maintenance of roads, bridges, and tunnels to ensure safe and effective transportation of people and goods. The book *Highway Engineering* includes the main topics and the basic principles of highway engineering and provides the full scope of current information necessary for effective and cost-conscious contemporary highway. The book reflects new engineering and building developments, the most current design methods, as well as the latest industry standards and policies. This book provides a comprehensive overview of significant characteristics for highway engineering. It highlights recent advancements, requirements, and improvements and details the latest techniques in the global market. *Highway Engineering* contains a collection of the latest research developments on highway engineering. This book comprehensively covers the basic theory and practice in sufficient depth to provide a solid grounding to highway engineers. This book helps readers maximize effectiveness in all facets of highway engineering. This professional book as a credible source and a valuable reference can be very applicable and useful for all professors, researchers, engineers, practicing professionals, trainee practitioners, students, and others interested in highway projects.

The book *Highway Engineering* consists of six chapters.

Chapter 1 represents a solution algorithm for optimizing traffic signal timings in urban road networks by considering reserve capacity with an equity constraint. It is well known that the variation of signal timings in a road network may cause an inequity issue with regard to the travel costs of road users traveling between origin-destination (O-D) pairs. That is, the users may be influenced differently by changing traffic signal timings. In this context, the bilevel programming model is proposed for finding reserve capacity for signalized road networks by taking into account the equity issue. In the upper level, the reserve capacity is maximized with an equity constraint, while deterministic user equilibrium problem is dealt in the lower level. In order to solve the proposed model, a heuristic solution algorithm based on harmony search combined with a penalty function approach is developed. The application of the proposed model is illustrated, for example, road network taken from literature.

In **Chapter 2**, transportation-architecture-people-focused triple, urban transportation, design of transportation systems, pedestrian-oriented design, and pedestrian-walkable spaces will be emphasized. However, by analyzing the effects of pedestrian dynamics, transport systems aim to present a mechanism with an improved model that will define dynamics for the first time to explain the processes underlying design decisions. Four hundred and twelve healthy volunteers were selected from 18 to 65 years of age. First, three-dimensional (3D) virtual city is designed to understand the experiences of the pedestrians. Later, it was provided to navigate the three-dimensional virtual glasses in the city where the broadcasts were designed. During this navigation, pedestrian dynamics were observed, and spaces where pedestrians cautioned were identified. Following this determination, the "attractive" locations will be shown on the macroscale, and with the eye-tracking method, it has been determined why "attractive" spaces are "attractive." In both virtual city navigation and analysis with eye-tracking technology, the cognitive activities of the broadcasts were tested with electroencephalogram (EEG). This approach will in general bridge an empirical and theoretical link between transport and architectural literature in understanding pedestrian movements/behaviors, combining architectural-pedestrian interactions with transport research. However, by analyzing the way-finding behavior, the study interprets its effects on the spatial area.

Cargo security during road transportation particularly presents a current topic in the context of transport safety. One of the key factors influencing the magnitude of impact (acceleration coefficients) during transportation is the quality of the road networks. Acceleration coefficient values directly affect the rate of inertia forces influencing the cargo. Given that the inertia force magnitudes (acceleration coefficients) are not known prior to commencing the actual transport, acceleration coefficient values known from regulations or otherwise (for example, empirically) certainly established values must be used. Values of acceleration coefficients were established in EN 12195-1, a regulation typically used within the European Union.

Chapter 3 covers the approaches of the above mentioned standard and provides comparison of acceleration coefficients established through regulations with those measured. Data (coefficient acceleration) from both highway transport and unpaved roads (in off-road conditions) were measured and statistically processed for comparison purposes. The transportation model presented subsequently demonstrates differences in the magnitude of inertia forces using three sets of data—acceleration coefficients obtained from the standard, from highway transport, and from off-road transport. At the same time, these secured cargoes were set into an insufficient context, where unsuitable or insufficient security of the cargo represents one of the significant risks in the occurrence of an accident. From the user's (drivers) point of view, a road is a unique linear infrastructure although it is clear for the driver that such a linear infrastructure consists of an open terrain (open roads) and occasionally a closed environment (tunnels). Since the two environments present different safety issues, the related analyses are usually conducted on the field by different experts; those conclusions are sometimes not well interlinked and harmonized, leading to safety gaps particularly, but not only, in the transition areas. Joint safety analyses conducted at the same time by a group of road safety and tunnel safety experts can fill such a gap and increase the safety level of the whole infrastructure. During the year 2016, an international group of road safety experts and tunnel experts visited five road sections with open roads and tunnel in Europe and performed joint safety analyses together with the infrastructure managers.

In **Chapter 4**, the above mentioned analyses were conducted according to a predefined experimental procedure to check the effectiveness of the joint analyses with respect to the usual ones. The key results are that joint safety operations in tunnels and open roads are possible and extremely useful: their cost can be very low when well planned. Most mega project infrastructure such as highway construction requires an enormous amount of cost. This situation might be a problem to developing countries that have limited national budget plan. On the other hand, the capability to transfer the funding of infrastructure depends on the project attractiveness to the private sector. The evaluation to involve in the project should be supported by a significant value of money from the business perspective. Optimum feasibility plays a vital role in bridging the partnership between the government and private interests.

Chapter 5 uses a case study of highway project development of Trans-Sumatra Highway Road (TSHR) that spans about 2,700 km to elaborate on how project feasibility can be improved through creative method. Value engineering (VE) approach as systematic ways to generate innovative ideas by combining multidiscipline background perspective is used. The study shows that additional functions for the project include integration with motorbike toll road, dry port, and railway line. Others also include tourism, fiber optic, and service area development. The innovative ideas have contributed to the significant increase in the internal rate of return (IRR) to the project from 7.79% to 12.76%. The chapter also formulates institutional scheme through build-operate-share-transfer (BOST), which administers government role in the project development.

Chapter 6 discusses thickness and elastic modulus of each pavement component, which then become input to elasticity analysis or finite element computation to calculate the service life of the highway. As an application was multichannel analysis of surface wave (MASW) measurement on a highway in North Jakarta. The measurement was carried out on highway traffic in Jakarta. The street was heading to Tanjung Priok, the port of Jakarta. The number of stacking was 10 to overcome traffic noise. The result of MASW measurement in terms of elastic modulus and thicknesses of pavement layer becomes an important input of finite element analysis to compute fatigue damage of pavement components.

Dr. Hamid Yaghoubi
Director of Iran Maglev Technology (IMT)
Iran

Chapter 1

Reserve Capacity Model for Optimizing Traffic Signal Timings with an Equity Constraint

Ozgur Baskan and Cenk Ozan

Additional information is available at the end of the chapter

http://dx.doi.org/10.5772/intechopen.70883

Abstract

This paper represents a solution algorithm for optimizing traffic signal timings in urban road networks by considering reserve capacity with an equity constraint. It is well known that the variation of signal timings in a road network may cause an inequity issue with regard to the travel costs of road users travelling between origin-destination (O-D) pairs. That is, the users may be influenced differently by changing traffic signal timings. In this context, the bilevel programming model is proposed for finding reserve capacity for signalized road networks by taking into account the equity issue. In the upper level, the reserve capacity is maximized with an equity constraint, whereas deterministic user equilibrium problem is dealt in the lower level. In order to solve the proposed model, a heuristic solution algorithm based on harmony search combined with a penalty function approach is developed. The application of the proposed model is illustrated for an example road network taken from a literature.

Keywords: reserve capacity, equity constraint, penalty function, bilevel programming, harmony search

1. Introduction

For decades, the term reserve capacity has been utilized for optimizing traffic signal timings as an objective. As known, it has been usually applied to signalized intersections and is defined as demand multiplier for existing origin-destination (O-D) demand matrix by considering cycle time, saturation capacity, green timings, and other corresponding constraints. Up to now, many researchers have carried out several works on this concept. The concept of reserve capacity was firstly applied by Webster and Cobbe [1]. They used the explicit formulation and calculated the reserve capacity of a signalized junction. Afterward, Allsop [2] formulated a linear program to calculate it for more complex signalized intersections. This approach proposed by Allsop [2] has been improved by using different saturation flows between stages by

Yagar [3, 4]. Wong [5] used the reserve capacity model to investigate the priority junctions and roundabouts. Wong and Yang [6] enhanced the reserve capacity model to a signalized road network with user equilibrium assignment. In their study, it was investigated that how the existing O-D demand matrix may be enlarged by providing that the link flows are operated under their capacities. Ziyou and Yifan [7] used two different objectives such as the reserve capacity and the link capacity expansion. For this purpose, they developed multiobjective solution algorithm, which aims to maximize the reserve capacity and to minimize the investment costs of link capacity expansions by considering signal timings. Ge et al. [8] provided a new approach to investigate how the reserve capacity is influenced by information of the road users. Results emphasized that the increase on the reserve capacity is not affected subsequently with the increase in information of the road users. Ceylan and Bell [9] developed a novel approach to calculate reserve capacity of a signalized network. In their study, optimum values of offsets, common cycle time, and the green times are investigated using genetic algorithm by taking into account stochastic user equilibrium link flows. Chen et al. [10] defined a new index related to capacity and reliability to determine demand multiplier for a signalized road network. The reserve capacity model is improved by considering the new developed index using a bilevel programming (BP) model. Chiou [11] proposed a new approach to find the demand multiplier in the context of reserve capacity with equilibrium constraints. Miandoabchi and Farahani [12] proposed an integrated model, which combined the problem of arrangement of lane directions and lane additions in urban road networks in terms of the reserve capacity. Chiou [13] developed a bilevel programming model to maximize the network's capacity. A hybrid-search heuristic method has been used to obtain maximum capacity of the road network by minimizing the total delay in terms of signal timings. The method has been applied to signalized road test networks. The results show that the developed algorithm outperforms the classical methods in terms of minimizing the delay. Wang et al. [14] proposed two bilevel programming models: one aims to maximize the reserve capacity in the network by using stochastic user equilibrium link flows, whereas the other one deals with maximizing reserve capacity by considering continuous network design problem. The significant result of their work is that the reserve capacity cannot always be increased by informing of the road users in the network. Xiao et al. [15] developed a reserve capacity model based on zone level to obtain maximum increase in travel demand and to investigate how the use of maximum capacity in the network influences the land-use development. For this purpose, the genetic algorithm has been used to solve the bilevel programming model and the proposed model is applied to a small numerical test network. The results reveal that the proposed reserve capacity model can be used for land-use planning in urban road networks. Recently, Han and Cheng [16] aimed to maximize the reserve capacity with a tradable credit scheme in the context of the stochastic user equilibrium. Numerical examples showed the advantage of the use of tradable credit scheme in maximizing reserve capacity in a given road network.

A short review of the literature shows that the studies for optimizing traffic signal timings within the context of reserve capacity ignored a significant issue. As known, the equilibrium travel costs between O-D pairs are influenced positively or negatively after changing the traffic signal timings in a road network. This may lead to different results for the road users traveling between O-D pairs. Consequently, the inequity problem is generated for the different road users travelling among O-D pairs on a road network. Therefore, the equity issue should be considered to implement more equitable signal control by the local authorities in order to prevent its negative effects on some road users traveling between O-D pairs. For decades, equity issue is started taking into account by researchers for solving some transportation problems. For example, Meng and Yang [17] examined the benefit of distribution between the road users and the inequity problem related to the continuous network design problem with regard to travel costs of users before and after applying a project related to network design. Yang and Zhang [18] proposed a bilevel programming model for the toll-design problem in a given road network by considering the equity constraint.

In the light of overview of literature, this study aims to improve the literature by considering the equity issue and the term of reserve capacity in the signal control problem, in order to provide equitable traffic signal timings on a given road network. A bilevel programming model with an equity constraint on the upper level is developed for determining the highest demand multiplier by setting traffic signal timings on a road network. We use the penalty function method to transfer the equity constraint as a penalty term to upper level objective function and then use harmony search (HS) method to solve it. In the lower level, the user equilibrium (UE) assignment in which the road users follow the user equilibrium principles of Wardrop is carried out using Frank-Wolfe (FW) algorithm proposed by Frank and Wolfe [19]. The rest of the paper is organized as follows: in the next section, an example is given to show the inequity problem occurred after changing the traffic signal timings. Section 3 presents the bilevel-heuristic solution algorithm with an equity constraint. In addition, the penalty function approach is given by incorporating HS approach. In Section 4, the numerical application of proposed model is presented. Conclusions and future studies are remarked in Section 5.

2. Inequity issue in the signal control problem

Consider a simple road network to show the inequity issue arises by changing the traffic signal timings. The road network has two O-D pairs (1→4, 2→4) as can be seen in **Figure 1**. The travel demands between the given O-D pairs are $q_{14}=450$ and $q_{24}=350$ vehicles per hour, respectively. It is assumed that dashed intersection is signalized as shown in **Figure 1**. The travel cost function used in the example road network is

$$t_a(x_a) = t_a^0 + \frac{x_a}{C_a} + (1 - \sigma_a)\Psi \tag{1}$$

where x_a is the flow on link a, t_a^0 is free-flow travel time, C_a is the capacity of link a, Ψ is the cycle time, σ_a is the green-time ratio, and $t_a(x_a)$ is travel time for link a. The parameters for the example road network are presented in **Table 1**.

It is supposed that the cycle time is 80 s for the signalized intersection numbered 3 in **Figure 1**. Let σ be the ratio of effective green time to cycle time on link 2, and it is taken as 0.5 for current situation. In this case, the green-time ratio on link 3 is 0.5. It is assumed that the behavior of the road users' route choice follows the UE conditions. In this case, the equilibrium travel costs between O-D pairs can be easily obtained using the FW algorithm as follows:

$$Z_{14} = 3.67, Z_{24} = 3.75 \quad (2)$$

where Z indicates the equilibrium travel cost for O-D pair w in W on the example road network. Suppose the local authority of road network has decided to decrease the green-time ratio on link 2 to 0.30. After applying of this arrangement, equilibrium travel costs between O-D pairs will be

$$Z_{14} = 3.59, Z_{24} = 4.10 \quad (3)$$

Eq. (3) implies that after decreasing the green-time ratio on link 2, the equilibrium travel cost from origin 1 to destination 4 is decreased and the travel cost from origin 2 to destination 4 is increased. This result means that the road users travelling between O-D pair (2-4) cannot get any benefit from this implementation. Consequently, the inequity problem exists among the road users travelling between O-D pairs. To minimize such negative effects of signal timing setting on a given road network, equity issue should be taken into account. For this purpose, Eq. (4) is proposed to show the level of changing equilibrium travel cost of each O-D pair before and after adjusting the signal timing.

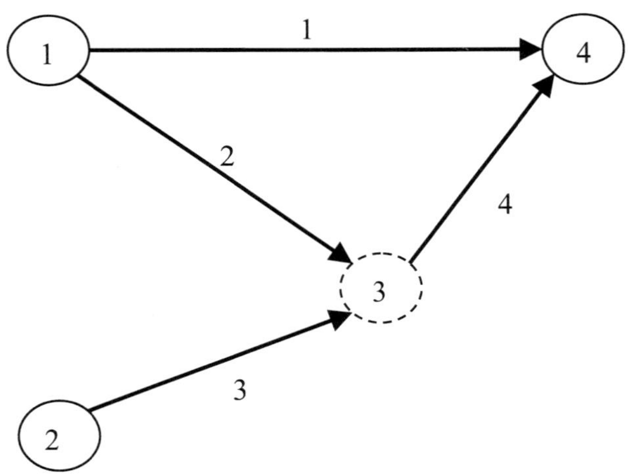

Figure 1. A simple road network.

Link no.	t_a^0	C_a	σ_a
1	3.5	1500	–
2	1.5	1200	σ
3	1.5	1500	$1-\sigma$
4	1.0	1500	–

Table 1. Parameters of travel cost function.

$$\max_{w \in W} \left| \frac{\overline{Z}_w - Z_w}{Z_w} \right| \le \beta \tag{4}$$

where $\overline{Z}_w$ and Z_w are defined equilibrium travel costs between O-D pair w in W before and after adjusting the traffic signal timing, and the parameter β is used to characterize the level of equity. This equation ensures that the variation of equilibrium travel cost of each O-D pair after adjusting traffic signal timing does not exceed a predetermined value.

3. Formulation of the problem

Let us assume that a road network has a set of links, A and a set of intersections, N. Its existing O-D demand matrix is represented as $\mathbf{q}$. Assume that a factor ξ is used to increase the existing O-D matrix, and thus the increased O-D matrix can be represented as $\xi\mathbf{q}$. Link flow $\mathbf{x}$ can be determined by considering the demand multiplier ξ and the green-time ratio σ for signalized intersections on a given road network. Thus, the objective function can be written as given:

$$\max_{\xi, \sigma} u = \xi \tag{5}$$

subject to

$$\max_{w \in W} \left| \frac{\overline{Z}_w(\mathbf{x}(\xi, \boldsymbol{\sigma})) - Z_w}{Z_w} \right| \le \beta \tag{6}$$

$$x_a(\xi, \sigma_\alpha) \le p_a C_a(\sigma_\alpha, s_a) \tag{7}$$

$$\mathbf{G}_i \boldsymbol{\sigma}_i \ge \mathbf{b}_i, \ \forall i \in I \tag{8}$$

Eq. (6) represents the constraint of equity and Eq. (7) is related to the link capacity constraint in which p_a represents degree of saturation for link a and s_a is the saturation capacity on link a. In Eq. (8), $\boldsymbol{\sigma}_i$ is a vector of green-time ratio for intersection $i \in I$, and matrix $\mathbf{G}_i$ and vector $\mathbf{b}_i$ are variables, which depend on the stage configuration of a given signalized intersection (see for details [20]). The equilibrium link flows, $x_a(\xi, \sigma_a)$, and corresponding travel costs for O-D pairs, $\overline{Z}_w(\mathbf{x}(\xi, \boldsymbol{\sigma}))$, are obtained by solving user equilibrium traffic assignment problem given below:

$$\min_x \sum_{a \in A} \int_0^{x_a} t_a^*(w, \sigma_a) dw \tag{9}$$

subject to

$$\sum_{k \in K} f_k^{rs} = \xi D_{rs} \quad \forall r \in R, \ s \in S, \ k \in K_{rs} \tag{10}$$

$$x_a = \sum_{rs} \sum_{k \in K_{rs}} f_k^{rs} \delta_{a,k}^{rs} \quad \forall r \in R, \ s \in S, \ a \in A, \ k \in K_{rs} \tag{11}$$

$$f_k^{rs} \geq 0 \quad \forall r \in R, \ \ s \in S, \ \ k \in K_{rs} \tag{12}$$

where t_a^* is the travel cost of all users on link $a \in A$. Eq. (10) represents that the sum of the route flows between O-D pair, *r-s*, has to be demand between same O-D pair. Eq. (11) shows that the flow on a link is to be the sum of the route flows which use this link. Eq. (12) is related to the non-negativity. Since the UE traffic assignment problem is convex, there are different methods, which are able to solve this problem efficiently. One of them is the Frank-Wolfe (FW) method. Indeed, this method is developed for quadratic optimization problems but it is also suitable for obtaining equilibrium link flows on transportation road networks [21, 22]. Therefore, the equilibrium link flows depended on the travel costs between O-D pairs and demand multiplier are calculated by using the FW method in this study.

In order to represent travel costs of users on link *a*, we suppose that it consists of two components such as flow-dependent running time and a stop delay arises from signal control at any intersection. This definition can be written as

$$t_a^* = \alpha_1 t_a(x_a) + \alpha_2(1 - \sigma_a)\Psi, \quad \forall a \in A \tag{13}$$

where t_a is the travel time on link $a \in A$, α_1 and α_2 are the values of travel time and stop delay, respectively. The right side of the Eq. (13) represents the stop delay occurred due to signalized intersection. If the considered intersection is not signalized, this term will be zero. It means that travel cost of all users on link *a* equals to its travel time. The link travel time $t_a(x_a)$ can be determined by using Bureau of Public Roads function as given below:

$$t_a(x_a) = t_a^0 \left[1 + 0.5 \times \left(\frac{x_a}{C_a} \right)^2 \right] \tag{14}$$

where t_a^0 is the free-flow travel time on link $a \in A$. Considering the reserve capacity maximization problem, the objective function given in Eq. (5) should be evaluated as minimization problem in keeping with the principles of the HS algorithm. Moreover, the constraints of Eq. (5) should be incorporated into the objective function by using the penalty function approach.

$$\min_{\xi, \boldsymbol{\sigma}} \overline{u} = \frac{1}{u} + \theta \left[\begin{array}{l} \max\left(\max\limits_{w \in W} \left| \dfrac{\overline{Z}_w(\mathbf{x}(\xi, \boldsymbol{\sigma})) - Z_w}{Z_w} \right| - \beta, 0 \right) \\ + \sum\limits_{a \in A} \max\left(x_a - p_a C_a, 0\right) + \sum\limits_{i \in I} \max(\mathbf{b}_i - \mathbf{G}_i \boldsymbol{\sigma}_i, 0) \end{array} \right] \tag{15}$$

subject to Eqs. (9)–(12)

where the first term of Eq. (15) is the inverse of the function *u*, which is the maximization problem, its second term is the penalty function in which θ is used as constant of the penalty, and so $\overline{u}$ is the minimization problem which will be solved by using the HS algorithm.

3.1. Bilevel programming model

As known, the bilevel programming (BP) model used for the solution of complex engineering problems is NP-hard. Even if the BP model has some disadvantages due to its non-convexity, it would be beneficial for taking into consideration the connection between interrelated problems. It is obvious that the BP model has two levels such as upper and lower levels. Even though both upper and lower levels are convex separately, the convexity of the both levels together cannot be ensured. This can be evaluated as a major disadvantage of the BP model. So, the exact algorithms such as Branch-Bound method can be incapable for the solution of the problem. In the context of reserve capacity, the upper levels deals with finding the maximum value of capacity by considering traffic signal timings while user's reactions to this changing are determined in the lower level with regard to equilibrium link flows. This mutual interaction can be evaluated by using the BP model effectively. On the other side, there has been increasing trend for the use of meta-heuristic algorithms in solving the BP model in recent years. Therefore, we used the harmony search (HS) algorithm for the solution of the BP model in this study. The HS algorithm is developed by Geem et al. [23]. As known, the musical process performed in an orchestra in order to reach the best harmony is the source of inspiration of this algorithm. The members of an orchestra try to reach the best harmony by making practices, just as searching of the global solution of a given problem can be controlled by some procedures in the HS. The HS algorithm uses four parameters for the solution. The first one is the harmony memory size (HMS), which represents number of population in the harmony memory (HM). The second one is the harmony memory considering rate (HMCR), which is the probability of the use of the HM to diversify the population. The third one is the pitch adjusting rate (PAR), which represents whether the pitch adjustment is required or not; and lastly the number of improvisations (NI), which represent the number of iterations (see for details [24, 25]). The procedure of the HS algorithm is summarized within the context of the reserve capacity maximization in order to brevity of the chapter as following:

Step 0: Set the HS parameters (HMS, HMCR, PAR, and NI) and network related parameters. The values of parameters HMCR, PAR, and HMS vary between 0.70–0.95, 0.20–0.50, and 10–50, respectively, which are proposed in order to obtain satisfying performance by Geem [26]. Therefore, HMS, PAR, and HMCR are used as 10, 0.45, and 0.85 in this study, respectively.

Step 1: Populate the HM with green-time ratio σ_a for each link $a \in A$ (i.e., link connected with signalized intersections) and demand multiplier (ξ).

Step 1.1: Solve the lower level problem (i.e., UE traffic assignment problem) by considering the variables $(\boldsymbol{\sigma}, \xi)$ generated in Step 1. After obtaining UE link flows, the objective function value of each harmony vector is calculated by using Eq. (15).

Step 2: Considering memory consideration, a new vector is calculated. The first member of the new vector is selected by using the HMCR parameter from the HM with probability of HMCR. The rest members of the vector are selected in similar way. On the other hand, the probability of the (1-HMCR) is used in order to decide whether the member of the new vector is selected from the feasible solution space.

Step 3: This step is related to decide whether the adjustment of the pitch is required or not. For this purpose, the PAR parameter is used and it varies between 0 and 1. After memory consideration step, the members of the new vector are increased or decreased with the probability of PAR by considering *bw* which is an arbitrary bandwidth (see for details [23]). The use of the probability of (1-PAR) provides that the corresponding value of the member of the new vector remains unchanged.

Step 4: Solve the lower level problem for the new vector. Then the corresponding objective function value is calculated by using Eq. (15).

Step 5: All objective function values stored in the HM are sorted from the best to the worst, and if the objective function value of the new vector is better than the worst one in the HM, the new vector is included into the HM instead of the worst one.

Step 6: Termination criterion. The algorithm is ended, if the number of iterations reached its maximum value. Otherwise go to Step 2.

The flowchart of bilevel-heuristic solution algorithm based on the HS is given in **Figure 2**.

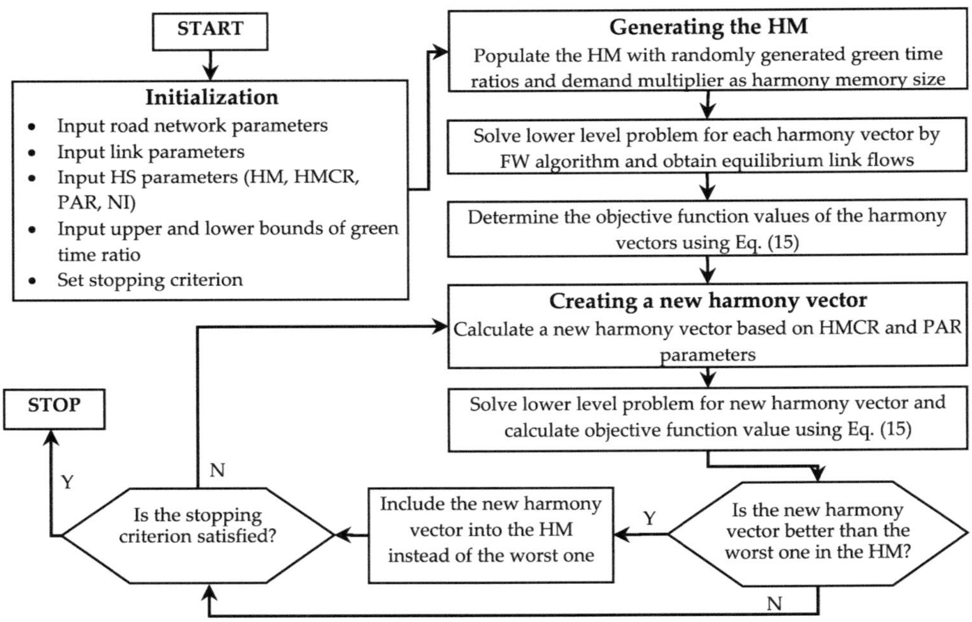

Figure 2. Flowchart of the bilevel-heuristic solution algorithm based on the HS.

4. Numerical application

The proposed bilevel-heuristic solution algorithm has been applied to example test network, which consists of 19 links, 13 nodes, and 4 O-D pairs as shown in **Figure 3**. The corresponding link parameters, such as free-flow travel time and saturation capacity, are given in **Table 2**.

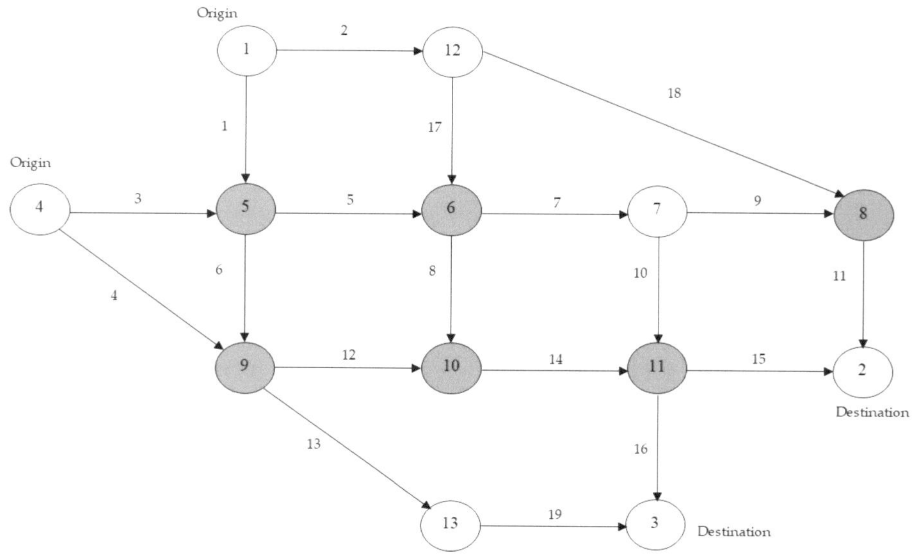

Figure 3. Example test network.

The number of link	Free-flow travel time t_a^0 (s)	Saturation capacity s_a (vehicle per hour)	Green-time ratio σ
1	270	5000	σ
2	270	3000	1
3	270	6000	$1-\sigma$
4	450	6000	$1-\sigma$
5	270	6000	$1-\sigma$
6	270	6000	σ
7	270	3000	1
8	270	6000	σ
9	270	3000	$1-\sigma$
10	270	5000	σ
11	270	3000	1
12	270	6000	$1-\sigma$
13	405	4000	1
14	270	6000	$1-\sigma$
15	270	2500	1
16	270	3000	1
17	270	3000	σ
18	540	4500	σ
19	270	4000	1

Table 2. Link parameters used in the test network.

The gray filled nodes are considered as signalized intersections as shown in **Figure 3**. Each of all signal-controlled intersections has two stages and their cycle time is assumed as 90 s. On the other hand, maximum and minimum values of the green-time ratio for each signalized intersection are 0.8 and 0.2, respectively. The O-D demands between nodes 1 and 2, nodes 1 and 3, nodes 4 and 2, and nodes 4 and 3 are 1500, 1000, 1000, and 1500 vehicle per hour, respectively. The maximum degree of saturation flow p_a is assumed as 0.9. Further assumptions are made on the values of travel time and stop delay and they are considered as $\alpha_1 = 20$ and $\alpha_2 = 40$ \$/hour [20].

The convergence graph of the bilevel-heuristic solution algorithm can be seen in **Figure 4**. It is assumed that the algorithm is terminated, when it reached the considered maximum number of iterations, namely 300. Taking into account of minimum and maximum values of green-time ratio, the proposed algorithm increases gradually the demand multiplier especially after 25 iterations. While the value of demand multiplier was about 1.1 at the first iteration, it reached the value of 1.55 with increasing of about 36% at the final iteration. It is obvious that this result has been achieved with taking into account the constraints of the link capacities, equity parameter, and signal timings. The equity parameter of β was fixed as 0.1 for this solution.

In this context, we have shown the equilibrium link flows after increasing total demand with multiplier of 1.55 as seen in the network given in **Figure 5**. While the total demand between four O-D pairs was 5000 before applying the demand multiplier to the network, it reached to 7777 vehicles per hour as shown in **Figure 5**. It should also be emphasized that there is no flow on the links numbered 8 and 9, which arises from the fundamental principle of the UE assignment. Another significant issue is that the links that carry lowest and highest flows are 17, 13, and 19.

It is pointed out that even then the links approaching to the signalized intersections in the network are operated under their capacities. This can be seen in **Figure 6**. In **Figure 6**, the gray bars represent capacities of the links approaching to signalized intersections, whereas the black bars show the equilibrium link flows. As can be seen in **Figure 6**, even though the equilibrium

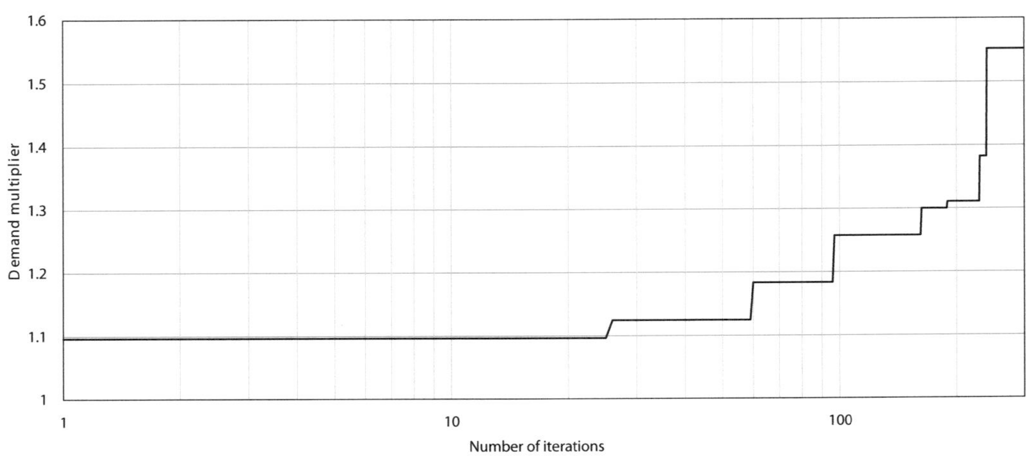

Figure 4. Convergence graph of the bilevel-heuristic solution algorithm.

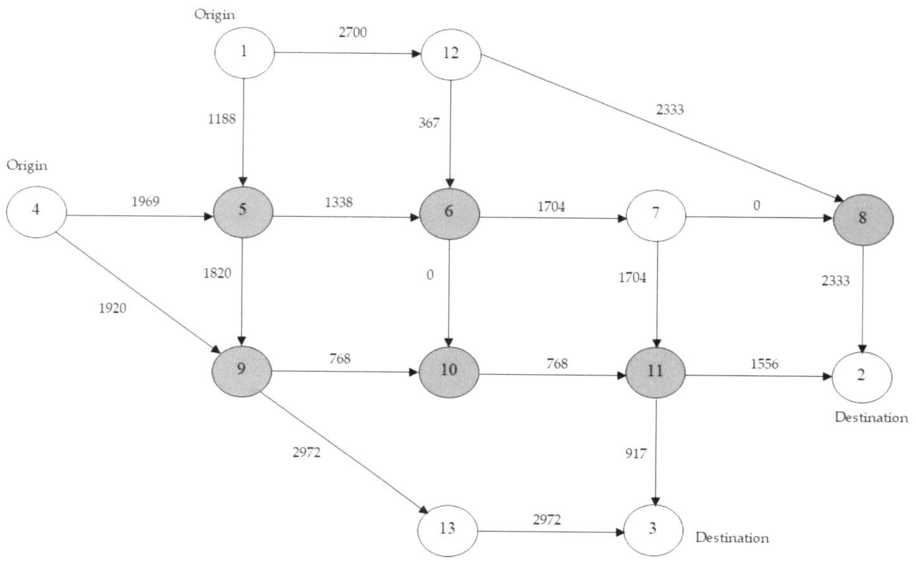

Figure 5. Equilibrium link flows after applying demand multiplier.

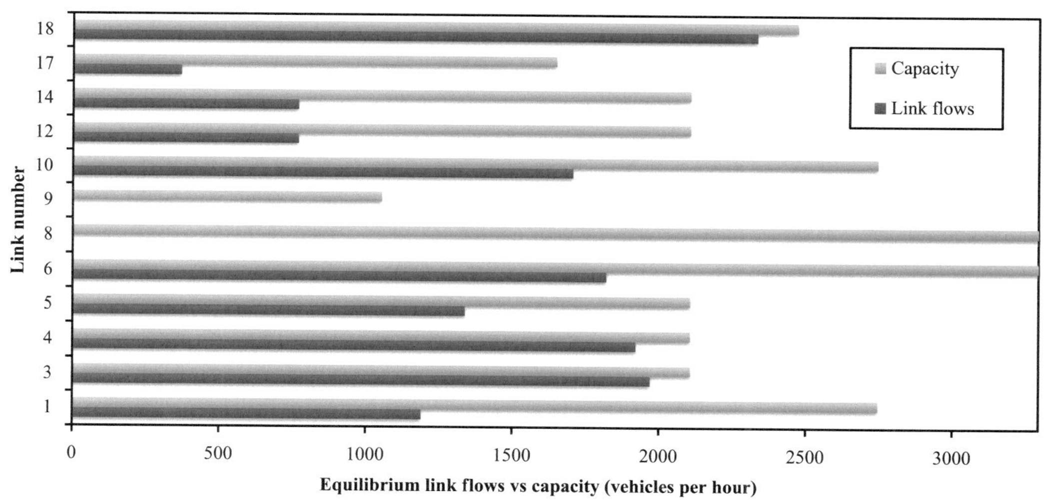

Figure 6. Comparison of equilibrium link flows and their capacities on signalized intersections.

flows on the links numbered 3, 4, and 18 are quite high, they continue to serve under their capacities.

The significant parameters of the links approaching to the signalized intersections are given in **Table 3**. The green-time ratios are also obtained after running the bilevel-heuristic solution algorithm under fixed value of parameter β as shown in **Table 3**. When we consider the signalized intersection numbered 5, which has two approaching links, namely 1 and 3, the green-time ratio has found 0.61 for the link 1 and 0.39 for the link 3. Since each of the signalized

The number of link	Equilibrium flow x_a (vehicles per hour)	Saturation capacity s_a (vehicles per hour)	Green-time ratio σ	Capacity $p_a C_a(\sigma_a, s_a)$ (vehicles per hour)
1	1188	5000	0.61	2745
3	1969	6000	0.39	2106
4	1920	6000	0.39	2106
5	1338	6000	0.39	2106
6	1820	6000	0.61	3294
8	0	6000	0.61	3294
9	0	3000	0.39	1053
10	1704	5000	0.61	2745
12	768	6000	0.39	2106
14	768	6000	0.39	2106
17	367	3000	0.61	1647
18	2333	4500	0.61	2471

Table 3. The parameters of the links approaching to signalized intersections for $\beta = 0.1$.

	$\beta = 0.05$	$\beta = 0.10$	$\beta = 0.15$	$\beta = 0.20$
Objective function value ($\overline{u}$)	0.672	0.643	0.640	0.640
Demand multiplier (ξ)	1.488	1.556	1.562	1.562
Green-time ratio (σ)	0.607	0.610	0.590	0.590

Table 4. Obtained results using different values of equity parameter β.

intersections has two stages, the sum of green-time ratios of two links is equal to one. In addition to this, we have shown also the capacities of links which are calculated taking into account degree of saturation, saturation capacity, and green-time ratio. Again, it can be seen that the flows of the considered links are operated under their capacities.

In order to test the effect of parameter β on the solution algorithm, it has been run for different values for β and the corresponding results are given in **Table 4**. While parameter β is increased from 0.05 to 0.20, the objective function values decrease in similar ratios. On the other hand, the demand multiplier increases since its value is inversely proportional to the objective function value. The reason for this is that the local authority pays less attention to the equity, so this leads to an increased travel demand in the network. Another significant result of this test is the objective function value and demand multiplier remain stable after a value of 0.15 of equity parameter β.

5. Conclusions

This study aims to optimize traffic signal timings in terms of reserve capacity by taking the equity constraint into account. In order to create more equitable networks, this phenomenon should be taken into consideration in optimizing traffic signal timings. Thus, the road users in the network are similarly taken advantage by changing traffic signal timings. For this purpose, we developed a bilevel-heuristic solution algorithm based on the HS. While the demand multiplier and green-time ratio are found in the upper level by considering the equity issue, capacity, and signal timings constraints, the UE link flows are determined in the lower level by using the FW algorithm. First, we showed the inequity problem for the users by changing traffic signal timings in a small test network. Results of this small test emphasized that the equity issue should be taken into account to benefit equally for all road users between O-D pairs by changing traffic signal timings. On the other hand, the middle-sized test network has been used in order to show the usability of the proposed solution algorithm for optimizing traffic signal timings with regard to the reserve capacity. It has been found that the proposed algorithm is able to maximize the reserve capacity for optimizing traffic signal timings by considering the equity issue. Results also show that the links in the test network are operated under their capacities, even the demand is increased about 1.5 times for fixed value of equity parameter. Another significant result of this study is that the proposed reserve capacity model is sensitive to the different values of equity parameter β. As the value of β is increased, the value of demand multiplier also increases. This arises from that the authority pays less attention to the equity in high values of β so this leads to an increased travel demand in the network.

Author details

Ozgur Baskan[1]* and Cenk Ozan[2]

*Address all correspondence to: obaskan@pau.edu.tr

1 Department of Civil Engineering, Engineering Faculty, Pamukkale University, Denizli, Turkey

2 Department of Civil Engineering, Engineering Faculty, Adnan Menderes University, Aydin, Turkey

References

[1] Webster FV, Cobbe BM. Traffic Signals. London: Ministry of Transport; 1966

[2] Allsop RE. Estimating the traffic capacity of a signalized road junction. Transport Research. 1972;**6**:245-255

[3] Yagar S. Capacity of a signalized road junction: critique and extensions. Transportation Research. 1974;**8**:137-147

[4] Yagar S. Addressing errors and omissions in papers on intersection capacity maximization. Transportation Research. 1985;**19B**:81-84

[5] Wong SC. Group-based optimization of signal timings using the TRANSYT traffic model. Transportation Research. 1996;**30B**:217-244

[6] Wong SC, Yang H. Reserve capacity of a signal-controlled road networks. Transportation Research. 1997;**31B**:397-402

[7] Ziyou G, Yifan S. A reserve capacity model of optimal signal control with user-equilibrium route choice. Transportation Research. 2002;**36B**:313-323

[8] Ge YE, Zhang HM, Lam WHK. Network reserve capacity under influence of traveller information. Journal of Transportation Engineering. 2003;**129**:262-270

[9] Ceylan H, Bell MGH. Reserve capacity for a road network under optimized fixed time traffic signal control. Journal of Intelligent Transportation Systems. 2004;**8**:87-99

[10] Chen A, Chootinan P, Wong SC. New reserve capacity model of signal-controlled road network. Transportation Research Record: Journal of the Transportation Research Board. 2006;**1964**:35-41

[11] Chiou SW. Reserve capacity of signal-controlled road network. Applied Mathematics and Computation. 2007;**190**:1602-1611

[12] Miandoabchi E, Farahani RZ. Optimizing reserve capacity of urban road networks in a discrete network design problem. Advances in Engineering Software. 2011;**42**:1041-1050

[13] Chiou SW. Optimal signal-setting for road network with maximum capacity. Information Sciences. 2014;**273**:287-303

[14] Wang J, Deng W, Zhao J. Road network reserve capacity with stochastic user equilibrium. Transport. 2015;**30**(1):103-116

[15] Xiao H, Gao J, Zou Z. Reserve capacity model based on variable demand for land-use development control. Transportation Planning and Technology. 2017;**40**(2):199-212

[16] Han F, Cheng L. Stochastic user equilibrium model with a tradable credit scheme and application in maximizing network reserve capacity. Engineering Optimization. 2017;**49**(4):549-564

[17] Meng Q, Yang H. Benefit distribution and equity in road network design. Transportation Research Part B. 2002;**36**:19-35

[18] Yang H, Zhang X. Multiclass network tool design problem with social and spatial equity constraints. Journal of Transportation Engineering. 2002;**128**:420-428

[19] Frank M, Wolfe P. An algorithm for quadratic programming. Naval Research Logistics Quarterly. 1956;**3**:95-110

[20] Li ZC, Ge XY. Traffic signal timing problems with environmental and equity considerations. Journal of Advanced Transportation. 2014;**48**:1066-1086

[21] Baskan O. Determining optimal link capacity expansions in road networks using Cuckoo search algorithm with Lévy flights. Journal of Applied Mathematics. 2013;**2013**:718015, 11 pages

[22] Baskan O. Harmony search algorithm for continuous network design problem with link capacity expansions. KSCE Journal of Civil Engineering. 2014;**18**(1):273-283

[23] Geem ZW, Kim JH, Loganathan GV. A new heuristic optimization algorithm: Harmony search. Simulation. 2001;**76**(2):60-68

[24] Dell'Orco M, Baskan O, Marinelli M. A Harmony Search algorithm approach for optimizing traffic signal timings. PROMET Traffic & Transportation. 2013;**25**(4):349-358

[25] Baskan O. An evaluation of heuristic methods for determining optimal link capacity expansions on road networks. International Journal of Transportation. 2014;**2**(1):77-94

[26] Geem ZW. Optimal Design of Water Distribution Networks using Harmony Search [thesis]. Seoul, Korea: Korea University; 2000

Chapter 2

Effects on the Design of Transport Systems of Pedestrian Dynamics

İlker Erkan

Additional information is available at the end of the chapter

http://dx.doi.org/10.5772/intechopen.70496

Abstract

In the study, transportation-architecture-people-focused triple, urban transportation, design of transportation systems, pedestrian-oriented design and pedestrian walkable spaces will be emphasized. However, by analyzing the effects of pedestrian dynamics, transport systems aim to present a mechanism with an improved model that will define dynamics for the first time to explain the processes underlying design decisions. Four hundred and twelve healthy volunteers were selected from 18–65 years of age. First, three-dimensional (3D) virtual city is designed to understand the experiences of the pedestrians. Later, it was provided to navigate the three-dimensional virtual glasses in the city where the broadcasts were designed. During this navigation, pedestrian dynamics were observed, and spaces where pedestrians cautioned were identified. Following this determination, the "attractive" locations will be shown on the macroscale, with the eye-tracking method, it has been in both virtual city navigation and analysis with eye-tracking technology, the cognitive activities of the broadcasts were tested with electroencephalogram (EEG). This approach will in general bridge an empirical and theoretical link between transport-architectural literature in understanding pedestrian movements/behaviors and combining architectural-pedestrian interactions with transport research. However, by analyzing the way-finding behavior, the study interprets its effects on the spatial area.

Keywords: pedestrian dynamics, transportation architecture, human behavior, pedestrian-based design, neuro-architecture science

1. Introduction

In recent years, pedestrian dynamics (movements and behaviors) have been linked to safety [1]; building evacuation [2]; transportation [3, 4] and physics, urban planning, and real-time practice [5]. Many different methods have been developed to understand pedestrian dynamics, among which the most commonly used are the "cellular automata model" [6, 7]; "The social force model" [8, 9] and the "agent-based model" [10, 11]. At the same time, the visual perceptions that surround the tracks during the course of the road respond positively/negatively [12, 13]. In other words, "visual perception" and "environmental stimuli" are two important influences that affect the dynamics of the pedestrian. Studies in the literature show that visually is related to pedestrian dynamics (movement and behavior). In many pedestrian models developed, because of the difficulties of real-time simulations of the pedestrians, the selection sets (directions) are divided into a limited number, then the desired walking direction of the pedestrians; the desired walking speed is determined by specific rules and mechanisms. However, in the models of previous pedestrian dynamics, the arc was regarded as the main dynamic, visual and other stimuli were ignored. Pedestrians are affected by many factors (stimulants) when walking in the city. Even if the objects are attractive enough, they can re-establish their standing in front of them. Although movement is a major influence on behavior, the interaction between the visual and the environment of the pedestrians is one of the inferences in the literature. However, in pedestrian dynamics, pedestrian-visual-environment relation, movement and/or behavior should be considered as directly affecting. One of the goals of this work is to develop a simulation model that can reveal realistic pedestrian dynamics and explain its mechanism. In this model, the effects of "architectural structures" on pedestrian dynamics, which are not mentioned in the literature, are examined. The developed model is called "attention-based visual motion (ABVM)."

This model tests the pedestrians, architectural visual attractors and then examines the characteristics of the attractor and the cognitive state of the stimulus to see whether the pre-defined roots of attention will dissipate. A three-dimensional (3D) small-scale city was designed for the model and different architectural stimuli were placed in the city. These different stimuli, distinctive façade designs and urban furnishings can attract direct attention of the pedestrians. It should not be forgotten that transportation design should benefit from simultaneous analysis of strategic and macroscopic cognitive states of pedestrians. The features of the stimuli (external attractors) in the 3D virtual city are defined and evaluated along with the necessary visually appealing characteristics of how these stimuli affect the pedestrian dynamics.

Existing studies in the literature have examined pedestrian dynamics in the context of transportation science. Besides, the analyses mentioned above have been made in the urban design scale, but the same methods have always been emphasized as methods. Moreover, none of these studies have examined in detail how pedestrian movements and/or behavior affect and shape the design. The study will evaluate this question in terms of design and cognition and will examine it both in the spaces where the pedestrians move constantly and in the space scale. The main aim of this study is to determine the effects of pedestrian dynamics (pedestrian movements and behaviors), visual and perceptual environment and architecture and to determine the architectural effects of these dynamics.

2. Material and method

In this study, it is aimed to determine the effects of pedestrian dynamics on visually and behaviorally structured environment and transportation systems and to determine the contribution/effect of architectural design. Depending on this purpose, the research will be carried out in three different parts, and three different research methods will be used in the study in stages and/or at the same time. In the study, this method, which is called the multiple analysis method, is a platform where three analytical techniques come together (Virtual Reality Glasses (3DVRG), eye tracking and electroencephalogram (EEG)) to interact with each other. This study generally analyzes pedestrian behavior in three different ways (eye tracking, 3DVR and EEG). Especially in the context of making sense of human movements, it is thought that the use of multiple analysis techniques together will be meaningful in the analysis of pedestrian dynamics.

2.1. Participants

Four hundred and twelve healthy volunteers were selected from 18–65 years of age. Two hundred and one man and 211 women have been participated in this experiment. Average age of man participants: 38,55. The average age of women is 35.66. Participants were informed about not taking any alcohol products 24 hours before the experiment. Since the participants should not use certain psychiatric medication, this question was asked during the election. It has been confirmed that there is no problem in the normal vision of the participants. In addition, participants should not use glasses or lenses, as the eye-tracking device used is not superior in feature, so only participants who have good vision and no apparatus are selected.

2.2. Instruments

An electroencephalogram (EEG) is a test used to evaluate the electrical activity in the brain. One 14-channel EEG was used for the experiment. These EEG channels have been used in this experiment: AF3, F7, F3, FC5, T7, P7, O1, O2, P8, T8, FC6, F4, F8 and AF4. Also, we used eye-tracking technology that eye-tracking data are collected using either a remote or head-mounted "eye tracker" connected to a computer. With eye-tracking technology, it is possible to determine where participants "look" and "for how long." However, 3D virtual reality glasses were used to make participants feel themselves in the virtual city environment. The recorded brain signals include artifacts such as muscle movement, eye blinking and so on. The signals were downsampled further to 128 Hz.

The eye-tracking device used in the experiment is binocular (double eye) tracking with data acquisition frequency of 120 Hz, eye movements can be recorded with 0.5° recording error. The refresh rate of the device was 120 Hz. The instrument can be individually calibrated in a very short time for each participant before the experiment. In the scope of the study, the eye-tracking measurements used were determined as "visual attention pattern" and "total fixation time." The visual angle of the screen used in the experimental phase is 32 × 29 and the visual angle of the stimuli is about 16° × 18°.

2.3. 3D virtual city

A virtual city is designed so that the pedestrians can feel themselves in the real environment. In this virtual city, there are elements such as walkways, buildings made according to different architectural styles, city furniture, landscapes and showcases. The movement of the pedestrians in the designed city was provided by the wii commander. The city has been visited in daylight. Because the 3D render in the daylight is more realistic and makes people more adaptable to the atmosphere. Examples of the designed 3d city are shown in **Figure 1**.

People who are seen in the city figure also act in the experiment. However, the vehicles act like people, which make the work more realistic.

2.4. Procedure

The total time for the entire experiment was set at 450 seconds. Three hundred seconds of this is a city tour. In the city tour, this part of experiment was conducted using 3D glasses. Participants passed the rest phase after 300 seconds. During this time, the software designed for this study identified "spaces explored for more than 10 seconds." These are called "attractive spaces." The participants were then again tested with eye tracking apparatus and maintained for 150 seconds. The experiment was terminated and removed from the participant list in order to feel symptoms such as dizziness and nausea during use of the virtual 3D glasses.

Participants are equipped with an EEG device and a virtual reality glasses during navigation. In order to be able to measure the net responses of the pedestrians, it was ensured that the EEG records were abstracted from different stimuli. The areas where they have been standing on the tracks are determined by video recordings. That is, if the pedestrians are too

Figure 1. 3D urban area.

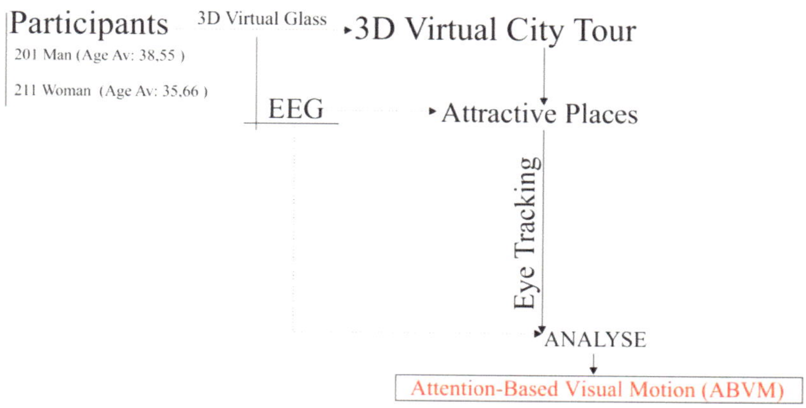

Figure 2. Experiment flowchart.

much to witness on a space (building front, walkways, landscaping, etc.), the system records the attractive space. Therefore, it will be possible to have information about why it stays so long and to be able to conduct a detailed examination. In detail, the test scheme is shown in **Figure 2**.

3. Data analyze

Many features are used in the literature to classify EEG data in brain-computer interface (BCI) for various mental tasks. Some of these are: band powers [14], power spectral density [15], autoregressive and adaptive autoregressive parameters [16], time-frequency features [17–20] and inverse model-based features [21–23]. In 1998, Norden E. Huang from NASA proposed a new signal analysis method named Hilbert-Huang transform (HHT) and it is applied to analyze nonlinear and non-stationary signals and was regarded as an important progress since the fast Fourier transform (FFT) [24]. Huang et al. [24] stated that the HHT method includes two steps. In the first step, the original data will be transformed into an intrinsic mode function (IMF), which satisfying the requirements of the Hilbert transform, by the method of empirical mode decomposition (EMD). In the second step, Hilbert transform method will be used on each order of IMF above to calculate its instantaneous frequency. All those results can be used to create an integrated time-frequency figure finally. In this study, Hilbert-Huang transformation, which is a classical time-frequency analysis method, was chosen.

According to [25, 26], for a real-valued g (t) time series, the Hilbert transform can be obtained as in Eq. (1).

$$g(t) = H\{g(t)\} = \frac{1}{\pi} P \int_{-\infty}^{\infty} \frac{g(\tau)}{t-\tau} \tag{1}$$

In Eq. (1), P represents the basic value of the complex integral Cauchy.

The Hilbert transformation given in Eq. (1) gives the analytic binary virtual part for the real data g(t) to obtain. However, to an original analytical signal, g (t) allows us to reach in Eq. (2).

$$g(t) = g(t) + \hat{g}(t) = G(t)\, e^{j\theta(t)} \quad (2)$$

The instantaneous frequency for the resulting signal, if the signal is a single component, $\theta(t)$ of the phase function is easily obtained by taking the derivative of the time in Eq. (3).

$$w(t) = \frac{d\theta(t)}{dt} \quad (3)$$

4. Experimental result

4.1. EEG data analyze

Given the features extracted from each participant's EEG data, the analysis attempted to determine which of the α, β, γ, $\theta \wedge \delta$ components of the EEG exhibited the highest mutual information with the objects that reflect the individual participant's preferences. During each of the different EEG channels, the normalized mutual information between the class tag and each of the four main EEG band powers was determined. Mutual information has been determined between 0 and 1. The detailed illustration of this is shown in **Figure 3**.

One of the most rewarding results of the study is a clear difference between the mutual information values obtained by symmetric channels in the left and right hemispheres in each of the EEG bands during city tour. It is thought that this difference shows asymmetrical activities in terms of strength of the EEG bands while making decisions about the places, the streets and their combination of the two, the triple combination preferred by the participants. Because some of the participants were concentrated on the street where the shops were concentrated, others were only directed toward the landscape areas. F3, F4 and O1 channels exhibited the

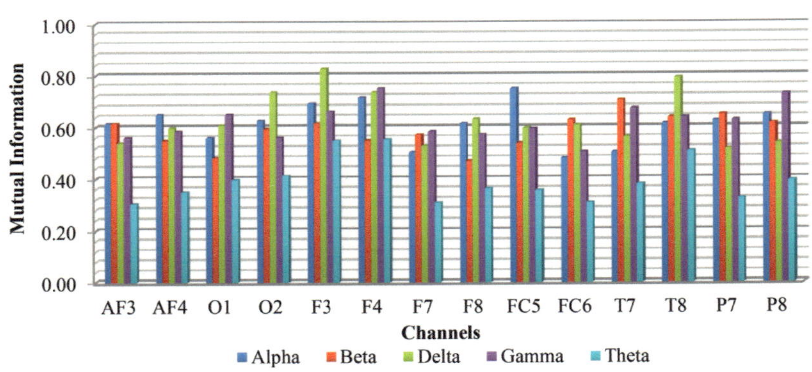

Figure 3. The normalized mutual information between the class tag and each of the four main EEG band powers.

highest mutual information in delta band. Delta oscillations are present in the background EEG activity not just in sleep, but in awake state and are thought to be generated by cortical networks [27]. Also, these oscillations play a significant role in large-scale cortical integration [28–30]. There are studies confirming the usability of the delta channel to distinguish the striking objects [31, 32]. It is noteworthy that the importance of the delta band during participant's way-finding preferences. In other words, serious subjects were taken especially in the channels F3 and F4 in the subjects who had difficulties in their preference of direction. During the way-finding of participants, information was observed in AF3, AF4, F3 and F4 channels, especially in those who have difficulty making decisions. The information of these channels indicates that this activity is predominantly in the frontal lobe.

Participants were observed to have differences in theta band when they inserted the eye-tracking device and questioned their preferences in detail. Theta band exhibited very high mutual information with the preference in the frontal region, especially in the channels F3, F4, FC6 and AF4. Theta band was especially associated with female participants, especially if there is a shopping mall between their preferences. That is, 78% of the female participants spent a long time in front of the stores. Therefore, mutual information was observed in the O2 and P8 regions in the case of detailed examination with eye-tracking technology. On the other hand, this information was observed only in the P8 region in some of the male participants. If the Theta band is thought to be associated with emotional situations [33, 34], it should not be surprising that this region differs in women. However, it was observed that they were standing in front of participants in the moment of passing Atatürk posters placed in certain shadows of the city. When they stopped, there was an increase in theta band in 83% of the participants.

Regarding the beta wave, the highest mutual information values are seen especially in the temporal and occipital regions in the city. In the eye-tracking phase, the frontal regions were again displayed with changes in the reference beta-spectrum. However, it is noteworthy that the changes in the F7 and F8 channels in brain activity during the eye-tracking phase, especially the façade's restoration feature. This can be interpreted as an indication of the changes in the emotional mode of people in the old built environment. Participants were encouraged to analyze that part of trying to pause in high-intensity horticultural gardens while touring landscape areas. A total of 394 people standing in colorful gardens recorded significant increases in beta bands in the occipital region. After leaving the colored garden and turning completely toward the green areas, the beta waves in the occipital area were recorded as declining. This may be an indication that the beta wave behaves in accordance with the color stimulus.

The highest spectral changes in F4 and P8 and O2 were observed, supporting the importance of both frontal and parietal regions in the gamma band.

According to the flow chart of the experiment, the participants are looking at a stimulus twice. In other words, they first look at the places that they see during the city tour and are determined to be interesting by using the eye-tracking device for the second time. The EEG device records in both stages. The areas that participants often identify as "attractive places" are listed below:

- Shopping center: The mall has designed modern façade. Participants have spent a long time viewing the broad front of the building.
- Historical street: While the participants are walking around the city center, there are five different streets to visit. Nearly all participants (96%) needed to visit the historic street.
- Atatürk bust and photographs: They spent a long time in front of the busts and photographs of Atatürk, the person with the highest degree of nationality.
- City square: The city square is the urban area where people can walk, sit and have living areas.
- Boutique stores: Stores are often attracted to women. On the other hand, posters and posters which are used as design items on the store also increase the attractiveness.
- Landscapes: Nearly all (98.7%) of the people have spent a large part of their time to see and even navigate the landscape area.

The relationship between the navigation in the modern areas and the historical sites has been examined. Beta and theta waves were used to help identify the relationship between these areas.

As shown in **Figure 4**, only the EEG is moving around the city (beta wave), showing mutual information in the back of the brain. However, in a more detailed analysis using eye-tracking

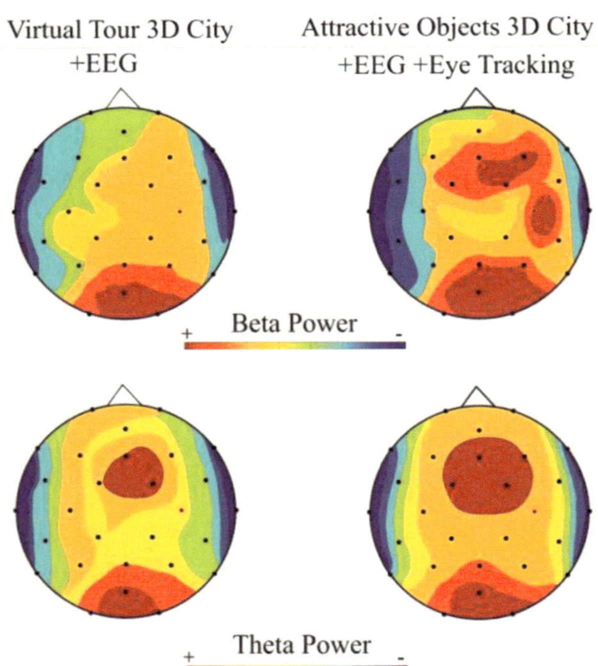

Figure 4. The difference between pedestrian brain map and pedestrians who make city tours with EEG device and those who examine both EEG and eye tracking.

technology, mutual information was observed in both the middle and frontal regions. But in theta wave, this information is not as remarkable as beta.

Here, there was an increase in brain activity, especially when walking through urban squares and green areas of the arcades. In addition, activities were recorded on both the beta and theta waves in areas where they prefer color landscapes rather than green landscapes and where there are colorful landscapes during this city tour.

The shops were placed in different locations on the walk and promenade routes of the designed 3D city. Some of these shops are ground floor and some are in the first floor. In this case, the interest of the pedestrians has been researched. When women's brain maps were examined, there was no difference between boutiques on the ground floor and boutiques on the first floor. So whether you are in the ground floor or the first floor is almost the same level of interest and attraction. However, both women and men spent more time in historical sites. Located on a modern street, the shopping center is focused on the spot after it has entered the view point, the shops that have been visited to the shopping center have been ignored and many have not been seen.

4.2. Eye-tracking data analyze

Areas where each participant has been standing for too long have been identified. In this case, the number is 3 in some participants and 8 in some participants. When you focus on a specific place for more than 10 seconds, the system records it automatically.

Figure 5 shows the number of places the total participants are focused on.

As shown in **Figure 5**, female participants are much more "affected by space" than men. More explicitly, the average locus of attractiveness of female participants throughout the city was 3.77; in males, this ratio remained at 2.29 (SD: 2.01).

Furthermore, It is also called the focus on focusing on one particular point [35, 36]. A factorial ANOVA was conducted to determine whether the total focus period on subjects of interest in the stimulus varied according to the content and participant's viewpoint ($p<0.05$).

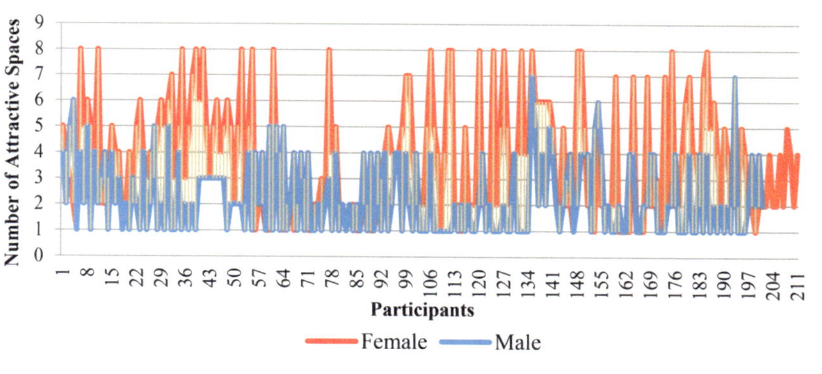

Figure 5. Number of spaces where each participant has held more than 10 seconds in the city tour.

Accordingly, the total duration of the focusing of the female participants during the excursion of the city to the details of the stimuli is longer than that of the male participants. This supports the view that the female participants are more interested in the stimuli (building façade, shops, etc.).

It is also possible to visualize the focus of participants' cognitive environment details with heat maps [37–39]. Heat maps show where participants look more at stimuli and are graded on a range of colors ranging from light green to red. The areas shown in green indicate that participants have fewer focuses, whereas the areas shown in red indicate that participants have more focus.

Figure 6 shows the details of some of the analyzed areas, which we can also call visual attention pattern.

Women Heat Map

Men Heat Map

Figure 6. Detailed display of visual attention pattern of some places.

In the section of the visual attention pattern examined separately for women and men, it is seen how much attention the female participants are paying attention to. As seen from the heat maps, males are usually concentrated on the "building entrance" directly. However, a large proportion of female participants were initially identified as "logos." We can see that this female cognitive behavior is not only in shopping mall structures but also in many store entrances.

5. Discussion

Findings include that the effects on the pedestrian moods of building facades, the green space on their health and the data about urban environments on their social interactions.

Behaviorally, people are often satisfied without going out of the green areas, but it has been found that the excitement understood from the analysis of the EEG records has increased in areas with more delicate landscape.

Naturally, it is thought that pedestrians prefer colored landscapes to green landscapes. The green areas that are the transportation axes to landscaping areas are preferred. In other words, the greening of the landscaping areas leads people to that point emotionally. This finding is very important for the design of walking paths. Because, if the designer wants to finish the hiking trail with landscaping areas, then designing the landscape along that trail is worthwhile in terms of giving people the feeling of directing them there.

However, in general, low spatial frequency components (large lines) have a high contrast, and high frequency components (small lines) have a low contrast. Simply, the scenes in the nature have lines that tend to exclude each other; so no lines appear in the picture when they are added. Therefore, in the nature, repetitive situations do not strike the eye; all the design elements are spread in a certain order. But this is not the case in urban scenes. Urban scenes violate the rule of nature: buildings tend to have regular, repetitive patterns due to the common use of design features such as unnatural landscapes, transportation systems and faulty restorations. There are studies [40, 41] in which the natural landscape structures positively affect the human brain. However, it is quite interesting that as a contribution to all studies, landscape areas also prefer "more colorful" areas of spring.

The shopping center is positioned within the city's modern buildings. An easy and central place has been chosen for vehicle access. However, a walkway has been connected in terms of reaching the arcades.

Because of this, the pedestrians have passed through many vehicles and reached the shopping center. This was detected by EEG, which increased the fear and anxiety mood during the transportation. From here, it can be thought that the pedestrians may be a problem to "pedestriate" from different vehicle routes until reaching such a central structure. Therefore, it would be appropriate to connect the pedestrian roads to the shopping centers and even to separate these roads from the roads.

From the architectural point of view, it is quite natural that the façades of the shopping center are closed and usually covered with advertisements. Therefore, it is noteworthy that in the analysis made by eye-tracking technology, the advertisement which is close to the entrance gate rather than the advertisement which is bigger is attracting attention. In other words, it is determined that the most focus point is seen from the eye-tracking heat map of the advertisement near the front entrance door. This shows that the closest ad to the frontline is the most striking when it comes to architectural design of the shopping center front.

The city square is an area where both women and men enjoy their time. However, the correct design of these areas is crucial both in terms of user satisfaction and design success. Another important feature of urban squares is that all transportation is united. In the 3D city that is designed, it is seen that the pedestrians move toward the city center. After reaching the center of the city, it was seen that the pedestrians were looking for a seating area or even heading toward outlying people.

The "green living areas" located in certain areas of the city center are located within the city. Almost all participants turn toward these areas. The artificial pools that are placed in the said space are the other preferred points. Therefore, it is seen that in urban centers, people turn to artificial pools first and secondly to green living areas and then to other living areas, starting from the orientation (moving toward that point) and eye-tracking data. Therefore, the satisfaction of broadcasting in the design of urban centers is much more important than in other places.

It has been determined that the springs have more time in the city than in other areas even if they are directed to the historical area designed in the city. At this point, it is particularly striking that women look more closely at the historical cities compared to men who travel more slowly. It is seen from the eye-tracking analysis that women are looking at the building facades while visiting the historical city, even examining some architectural items (beaks, beads, etc.) in detail. In addition to this, men's trips to historical cities have been determined from a general point of view. At the beginning of the modern constructions that started at the end of the designed 3D city, both women and men had right-lobe activities.

There are different boutiques that are designed and placed in different parts of the virtual city (by the side of the walkway, in the middle of modern buildings, next to the old city, etc.). These are small shops from different sectors such as footwear stores, electronic gadgets, books and stationery and music centers. It is observed that women spend more time in all kinds of boutiques than men and even spend a long time in front of the stores. On the other hand, in the direction of the data obtained from the eye-tracking technology, it was determined that women first looked at the brands of the magazines and later they looked for the entrance doors.

The point of interest is the similarity of the reaction with the eye-tracking technology of the female volunteers participating in the experiment from almost all ages. So, first they were interested in the magazine brand and then examined the entrance door details. In the EEG records, more dominant information was observed in women in the right brain region.

Therefore, aesthetic concern in the design of a woman's boutique keeps the front panel. Conversely, most of the male participants (85%) have not observed any movement in this region. I mean, when you go shopping for a boy, you get cheap, good quality, you care about your worries.

We had these results by measuring how the brain affected the images of natural and urban scenes. There are two ways to measure productivity; the first is to build simple computer models in the form of calculating the views of nerve cells. Any kind of design broken from the nature in the study leads to an increase in brain activities. It was observed that participants who spent time in areas not particularly affected by nature of urban areas, especially constructed from gummy concrete, glass, were active in both brain regions. In addition, a different fluctuation was observed especially in the frontal lobe compared to other lobes.

It may be thought that the pedestrians are seriously forcing the brain in designs that are completely human in nature and do not benefit from the slightest object of nature. In other words, such images make more effort to process the brain. One of the most striking aspects of the work is that it is irritating to distinguish nature from the design rule, and it has been detected both by cognitive and eye-tracking technology that artificial surroundings harm human beings. In addition, participants were found to move 1.4 times faster than digger sites in order to avoid such ultra-modern areas.

The worries and anxieties that arise in the fields of broadcasts reduce their satisfaction; thus, turning navigation into a negative direction. At the same time, it is evident that traveling in such spaces affects people emotionally.

It is remarkable that spatial studies of aesthetics are questioned and design is questioned [42, 43]. The role of design in different disciplines such as the effect of light design, the mathematical view of visual images [44], the research ability of design in terms of neuroscience [45], the meaning of building facades and human perception [46, 47].

However, in this study where the analysis and effect of transportation systems are discussed, it can be seen that while the transportation systems are designed, both the utilization of architectural discipline and the support of different disciplines such as brain science and cognitive science will facilitate the design of the systems and even turn them into a certain system.

6. Conclusion

Analyzing how the built environment influences your brain through blood-based research, the transport engineer or architects can provide you with the insight necessary to make healthier and more socially conscious designs.

Designing the system by following the behavior of the pedestrians and the user-oriented nature of this system will be appropriate in terms of future living spaces. It should be assumed that designs in which both transportation systems and architectural disciplines coexist are human-centered. Therefore, if the first focal point of the design is thought to be "human," then the human-oriented design logic will also be true. However, in order to be able to do this, deriving conclusions from the behavior and viewing of the pedestrians can open up both the strength of design and the correct design of the delivery systems. In this study, the areas in which emotions they feel when they navigate in the virtual environment are analyzed by EEG and then found "attractive" are analyzed by eye-tracking technology. As a result, it has been found that people are more likely to spend more time in areas where water and landscaping are enjoyed, which they like more about nature. It has been determined that people are more "comfortable" in such spaces. However, in the areas where modern structures exist, "uneasiness" and "anxiety" conditions have increased during the trips. These places are among the most important finds to see the bushes look good and do not focus.

The design should have areas that are not just about the look, but that can be functional as well as the look that they can experience, suitable for its purpose, capable of raising the quality of life, improving our health, social behavior and productivity.

Acknowledgements

Special thanks to Mesut Arslan (B.Sc.) for designing virtual city and animation.

Author details

İlker Erkan

Address all correspondence to: ilkererkan@sdu.edu.tr

Faculty of Architecture, Süleyman Demirel University, Isparta, Turkey

References

[1] EARPA (European Automotive Research Partners Association). Future Road Vehicle Research R and D Technology Roadmap. A Contribution to the Identification of Key Technologies for a Sustainable Development of European Road Transport. 2009. ISBN: 3-200-00017-1

[2] Shiwakoti, Nirajan, and Majid Sarvi. Enhancing the panic escape of crowd through architectural design. Transportation Research Part C: Emerging Technologies. 2013;**37**:260-267

[3] Davidich M, Geiss F, Mayer HG, Pfaffinger A, Royer C. Waiting zones for realistic modelling of pedestrian dynamics: A case study using two major German railway stations as examples. Transportation Research Part C: Emerging Technologies. 2013;**37**:210-222

[4] Duives DC, Daamen W, Hoogendoorn SP. State-of-the-art crowd motion simulation models. Transportation Research Part C: Emerging Technologies. 2013;**37**:193-209

[5] Dijkstra J, Timmermans HJP, Jessurun HJP. A multi-agent cellular automata system for visualising simulated pedestrian activity. In: Theory and Practical Issues on Cellular Automata. Springer; 2001; pp. 29-36, Editors: Stefania BandiniThomas Worsch

[6] Bandini S, Federici ML, Vizzari G. Situated cellular agents approach to crowd modeling and simulation. Cybernetics and Systems: An International Journal. 2007;**38**(7):729-753.

[7] Burstedde C, Klauck K, Schadschneider A, Zittartz J. Simulation of pedestrian dynamics using a two-dimensional cellular automaton. Physica A: Statistical Mechanics and its Applications. 2001;**295**(3):507-525

[8] Kwak, Jaeyoung, et al. Collective dynamics of pedestrians interacting with attractions. Physical Review E. 2013;**88**(6):062810

[9] Xu S, Duh HB-L. A simulation of bonding effects and their impacts on pedestrian dynamics. IEEE Transactions on Intelligent Transportation Systems 2010;**11**(1):153-161

[10] Bandini S, Manzoni S, Vizzari G. Modeling, simulating, and visualizing crowd dynamics with computational tools based on situated cellular agents. In: Pedestrian Behavior: Models, Data Collection and Applications. Emerald Group Publishing Limited. 2009. pp. 45-62, ISBN: 978-1-84855-750-5,Edited by: Harry Timmermans

[11] Granie M-A, Pannetier M, Gueho L. Developing a self-reporting method to measure pedestrian behaviors at all ages. Accident Analysis & Prevention 2013;**50**:830-839

[12] Zacharias J. Pedestrian behavior pedestrian behavior and perception in urban walking environments. CPL Bibliography. 2001;**16**(1):3-18

[13] Zacharias J. Exploratory spatial behaviour in real and virtual environments. Landscape and Urban Planning 2006;**78**(1):1-13

[14] Palaniappan, Ramaswamy. Brain computer interface design using band powers extracted during mental tasks. In: Neural Engineering Conference Proceedings. 2nd International IEEE EMBS Conference on 2005. p. 321-324, VA, USA, Publisher:IEEE Explore

[15] Carrier J, Land S, Buysse DJ, Kupfer DJ, Monk TH. The effects of age and gender on sleep EEG power spectral density in the middle years of life (ages 20-60 years old). Psychophysiology. 2001;**38**(2):232-242

[16] Gwin JT, Gramann K, Makeig S, Ferris DP. Removal of movement artifact from high-density EEG recorded during walking and running. Journal of Neurophysiology. 2010;**103**(6):3526-3534

[17] Nai-Jen, H., & Palaniappan, R.. Classification of mental tasks using fixed and adaptive autoregressive models of EEG signals. In Engineering in Medicine and Biology Society, 2004. IEMBS'04. 26th Annual International Conference of the IEEE 2004; Vol. 1, pp. 507-510

[18] Wang T, Deng J, He B. Classifying EEG-based motor imagery tasks by means of time–frequency synthesized spatial patterns. Clinical Neurophysiology. 2004;**115**(12):2744-2753

[19] Papandreou-Suppappola A, ed. Applications in Time-Frequency Signal Processing. CRC Press; 2002. p. 10, Taylor & Francis LLC

[20] Al-Fahoum AS, Al-Fraihat AA. Methods of EEG Signal Features Extraction Using Linear Analysis in Frequency and Time-Frequency Domains. ISRN Neuroscience; 2014, Editors: A. Grant, J. A. Hinojosa, and M. S. Oliveira, Hindawi Publishing Corporation

[21] Lotte F, Congedo M, Lécuyer A, Lamarche F, Arnaldi B. A review of classification algorithms for EEG-based brain–computer interfaces. Journal of Neural Engineering. 2007;**4**(2):R1

[22] Lotte F, Lécuyer A, Arnaldi B. FuRIA: An inverse solution based feature extraction algorithm using fuzzy set theory for brain–computer interfaces. IEEE Transactions on Signal Processing. 2009;**57**(8):3253-3263

[23] Tadel F, Baillet S, Mosher JC, Pantazis D, Leahy RM. Brainstorm: A user-friendly application for MEG/EEG analysis. Computational Intelligence and Neuroscience 2011;**2011**:8

[24] Huang, Manling, et al. Application and contrast in brain-computer interface Between hilbert-huang transform and wavelet transform. In: Young Computer Scientists, 2008. ICYCS 2008. The 9th International Conference for. IEEE, 2008. p. 1706-1710, Hunan, China, Publisher:IEEE Explore

[25] Huang NE, Attoh-Okine NO (eds.) The Hilbert-Huang Transform in Engineering. CRC Press; 2005, Taylor & Francis Group

[26] Huang NE. Hilbert-Huang Transform and Its Applications. Vol. 16. World Scientific Publishing Co. Pte. Ltd; 2014, ISBN: 978-981-4508-23-0

[27] Stefanics G, et al. Phase entrainment of human delta oscillations can mediate the effects of expectation on reaction speed. Journal of Neuroscience 2010;**30**(41):13578-13585

[28] Bruns A, Eckhorn R. Task-related coupling from high- to low-frequency signals among visual cortical areas in human subdural recordings. International Journal of Psychophysiology 2004;**51**:97-116

[29] Babiloni C, Brancucci A, Vecchio F, Arendt-Nielsen L, Chen AC, Rossini PM. Anticipation of somatosensory and motor events increases centro-parietal functional coupling: An EEG coherence study. Clinical Neurophysiology 2006;**117**:1000-1008

[30] Padilla ML, Wood RA, Hale LA, Knight RT. Lapses in a prefrontal-extra striate preparatory attention network predict mistakes. Journal of Cognitive Neuroscience 2006;**18**:1477-1487

[31] Knyazev GG. EEG delta oscillations as a correlate of basic homeostatic and motivational processes. Neuroscience & Biobehavioral Reviews. 2012;**36**(1):677-695

[32] Kurt P, Eroğlu K, Kuzgun TB, Güntekin B. The modulation of delta responses in the interaction of brightness and emotion. International Journal of Psychophysiology 2017;**112**:1-8

[33] Knyazev GG, Slobodskoj-Plusnin JY, Bocharov AV. Event-related delta and theta synchronization during explicit and implicit emotion processing. Neuroscience. 2009;**164**(4):1588-1600

[34] Bekkedal MY, Rossi J, Panksepp J. Human brain EEG indices of emotions: delineating responses to affective vocalizations by measuring frontal theta event-related synchronization. Neuroscience & Biobehavioral Reviews. 2011;**35**(9):1959-1970

[35] Granka LA, Joachims T, Gay G. Eye-tracking analysis of user behavior in WWW search. In: Proceedings of the 27th Annual International ACM SIGIR Conference on Research and Development in Information Retrieval. ACM; 2004. pp. 478-479, Sheffield, United Kingdom

[36] Cutrell E, Guan Z. What are you looking for? An eye-tracking study of information usage in web search. In: Proceedings of the SIGCHI Conference on Human Factors in Computing Systems. ACM; 2007. pp. 407-416, San Jose, California, USA

[37] Holmqvist K, et al. Eye Tracking: A Comprehensive Guide to Methods and Measures. Oxford University Press; 2011; ISBN:978-0-19969708-3

[38] Špakov O, Miniotas D. Visualization of eye gaze data using heat maps. Elektronika ir Elektrotechnika, 2015;**74**(2):55-58

[39] Djamasbi S, Siegel M, Tullis T. Visual hierarchy and viewing behavior: An eye tracking study. In: Human-Computer Interaction. Design and Development Approaches. 2011.

pp. 331-340, 14th International Conference, HCI International 2011, Orlando, FL, USA, Editor: Julie A. Jacko

[40] Barton J, Griffin M, Pretty J. Exercise-, nature-and socially interactive-based initiatives improve mood and self-esteem in the clinical population. Perspectives in Public Health 2012;**132**(2):89-96

[41] Le AT, Payne J, Clarke C, Kelly MA, Prudenziati F, Armsby E, Penacchio O, Wilkins AJ. Discomfort from urban scenes: Metabolic consequences. Landscape and Urban Planning, 2017;**160**:61-68

[42] Sullivan WC, Lovell ST. Improving the visual quality of commercial development at the rural–urban fringe. Landscape and Urban Planning. 2006;**77**(1-2):152-166

[43] Gjerde M. Visual evaluation of urban streetscapes: How do public preferences reconcile with those held by experts? Urban Design International. 2011;**16**(3):153-161

[44] Edelstein, E.A., Gramann, K., Schulze, J., Shamlo, N.B., Van Erp, E., Vankov, A., Makeig, S., Wolszon, L. and Macango, E.: Neural Responses during Navigation in the VirtualAided Design Laboratory: Brain Dynamics of Orientation in Architecturally AmbiguousSpace. Movement and Orientation in Built Environments: Evaluating Design Rationale and User Cognition, 2008, V (1), p. 35

[45] Shahroudi AA, Shabani H, Maboudi K, Lasgari R. Facades of Building Significantly Modulate EEG Signals of Brain Cortical Lobes, 2016 ANFA Conference. pp. 112-113, University of California, San Diego, Publiser: Aalborg University

[46] Essawy S, Kamel B, Samir M. Sacred buildings and brain performance: The effect of Sultan Hasan Mosque on brain waves of its users. Creative Space. 2014;**1**(2):123-141

[47] Kacha L, Matsumoto N, Mansouri A. Electrophysiological evaluation of perceived complexity in streetscapes. Journal of Asian Architecture and Building Engineering. 2015;**14**(3):585-592

Chapter 3

Analysis of Highway Acceleration in Regard to Cargo Security

Martin Vlkovský

Additional information is available at the end of the chapter

http://dx.doi.org/10.5772/intechopen.70829

Abstract

Cargo security during road transportation particularly presents a current topic in the context of transport safety. One of the key factors influencing the magnitude of impact (acceleration coefficients) during transportation is the quality of the road networks. Acceleration coefficient values directly affect the rate of inertia forces influencing the cargo. Given that the inertia force magnitudes (acceleration coefficients) are not known prior to commencing the actual transport, acceleration coefficient values known from regulations or otherwise (for example, empirically) certain established values must be used. Values of acceleration coefficients were established in EN 12195–1, a regulation typically used within the European Union. This chapter covers the approaches of this standard and provides comparison of acceleration coefficients established through regulations with those measured. Data (coefficient acceleration) from both highway transport and unpaved roads (in off-road conditions) were measured and statistically processed for comparison purposes. The transportation model presented subsequently demonstrates differences in the magnitude of inertia forces using three sets of data—acceleration coefficients obtained from the standard, from highway transport, and from off-road transport. At the same time, these secured cargoes were set into an insufficient context, where unsuitable or insufficient security of the cargo represents one of the significant risks in the occurrence of an accident.

Keywords: transportation, cargo security, transport model, acceleration coefficient, inertial force, accident rate

1. Introduction

The growing demand for cargo transport is related to the increase of transport performance in road transport. This is evident not only in Europe but also in the Czech Republic, which is a transit state for road haulers.

In 2015 in the Czech Republic, 438,906,000 tonnes of cargo was totally transported. Compared to 2010, it is more than a 23.3% increase (355,911,000 tonnes) [1].

Studies and statistics show that in the last decade, the gross weight of goods in containers handled in European ports increased by 30%. Many roads are already overloaded, because over three quarters of the total cargo volume (76.4%) in Europe is carried on roads [2]. The road infrastructure, which is highly loaded and often overloaded, is increasingly damaged, and annual maintenance often fails to provide the required quality of road infrastructure.

According to the data from the Road Transport Services Center (established by the Ministry of Transport of the Czech Republic), over half of vehicles are overloaded during weight checks. The number of weight checks surpasses 2000 per year [3].

Road quality (among others) directly affects the inertia force effect of cargo during transport. Generally, for a damaged road that has considerable unevenness (holes, potholes, etc.), higher values of acceleration coefficients can be assumed, which directly affects the value of inertia forces.

Load securing, also known as cargo securing, is the securing of cargo for transportation. According to the European Commission Transportation Department, it has been estimated that up to 25% of accidents involving trucks can be attributed to inadequate cargo securing [4].

From the point of view of the safety of cargo transport, one must consider the value of assumed inertia forces affecting the cargo and secure (fasten) the cargo with appropriate fastening means that correspond to the assumed inertia forces. The estimated quarter of traffic accidents, where an incorrect or insufficient load attachment is identified as a cause, are usually a combination of greater than the expected inertia forces and inappropriate attachment.

Determining the magnitude of the inertia for transport is possible using accelerometers and the appropriate calculation (e.g., using the relevant European standards). However, the size of the inertial forces must be known in advance and a suitable and sufficient method of fixing the cargo should be chosen accordingly. For this purpose, the empirically determined and statistically evaluated acceleration coefficients are used, which are part of EN 12195-1—Safety—Part 1: Calculation of securing forces, valid in the European Union.

The shortage is mainly at statistical processing, which is "averaged" for Europe. However, it is evident that the quality of the transport infrastructure and possibly also the quality of the various categories of roads differ significantly, not only when comparing several European countries but also within a given country. The Czech Republic is not an exception. Differences can be identified between repaired (upgraded) and older (not repaired) sections of the same road.

The aim of the chapter using the case study is to point out the deficiencies in the use of the abovementioned empirically determined and statistically processed values of acceleration coefficients. This thesis will subsequently be demonstrated within the framework of the performed transport model.

2. Acceleration and its impact on cargo securing

Acceleration, as well as vibrations, may adversely affect cargo transport. This is especially the case when, for some reason, cargo is detached and subsequently damaged. As a result of the detaching of the cargo, other technical elements such as fasteners (e.g., straps), transport means (e.g., pallets, containers), or the means of transport itself may be damaged. In some cases, injuries to the operator may occur. Such cases occur mainly in international traffic, where the goods are transported over long distances by more modes of transport (within multimodal transport).

A relatively self-contained area is also an incorrectly located center of gravity of the handling unit due to inappropriate loading—whether it is a container unit, a unit using an exchangeable extension, or a balancing of its own means of transport. In such cases, the appropriate handling and transport unit may be removed from the transport—most often in the case of air transport, where the requirement to place the center of gravity of handling and transport unit (container unit) is most strict.

A bigger problem, which may be caused by incorrect placement of the center of gravity of the handling and transport unit, is a particular problem during handling and transport in the case of container units designed for multimodal transport. In such cases, such a handling and transport unit (container) may not meet the requirements of the relevant international standard (ISO) [5], which states the basic load distribution condition in standardized ISO container type 1 (see **Figure 1**).

A simple condition relating to 60% of the weight of the load in the half of the container unit can be transformed, using the balance on the lever into the maximum deviation of the center of gravity of the container unit ±10% in the x-axis (see **Figure 1**). **Figure 1** shows a situation where the center of gravity in the x-axis is moved by the maximum possible deviation to the left, i.e., the position of the center of gravity of the x-axis container unit is 10%.

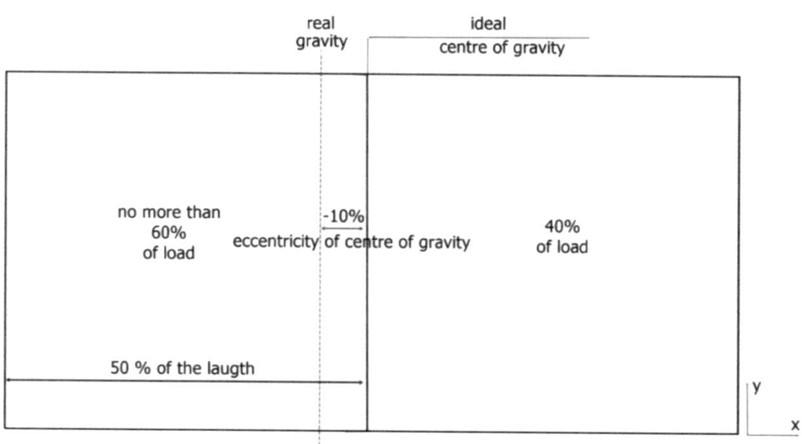

Figure 1. Load distribution in ISO containers. Source: modified from Refs. [5, 6].

The x-axis is the longitudinal axis according to the standard marking in relation to the movement of the means of transport (see **Figure 2**).

It is worth mentioning that much software is designed to load containers. The software contains conditions that ensure the permissible location of the center of gravity of the container unit and is usually supplemented by the same condition for the y-axis (according to the standard marking of the axis perpendicular to the x-axis, i.e., the axis of movement of the means of transport). With sensitive loads, this condition can be tightened and is available with certain software products.

Simply said, if the cargo of the handling and transport unit (or the direct means of transport) is properly distributed and secured, it is assumed that the abovementioned negatives (risks) will not occur. Obviously, there may be unexpected situations when cargo can be detached during handling, especially loading, unloading, and shipping.

The issue will be examined from the point of view of the possible cause of a traffic accident due to insufficient or improper attachment of the cargo at handling and transport unit. Cases where an accident or similar unexpected situation occurs will not be further investigated.

Determining the magnitude of the inertial forces stems from Newton's second law of motion, which defines the force as the product of mass—in this case, the mass of the load (m) and the acceleration (a). For the purposes of the transport model, the magnitude of the inertia force at given axis will be relevant, not its vector expression (direction):

$$F = m \cdot a \qquad (1)$$

Acceleration can be defined as corresponding acceleration coefficient (i.e., for the respective x-, y-, or z-axis) that is dimensionless in the context of the values of acceleration coefficients used, which are determined (empirically measured) normatively (see EN 12195-1), multiplied by gravitational acceleration (g):

$$a = c_{x,y,z} \cdot g \qquad (2)$$

Although gravity acceleration ranges from 9780 ms^{-2} on the equator to 9832 ms^{-2}, which corresponds to the Earth's pole [8], for the needs of multimodal transport, logistics and

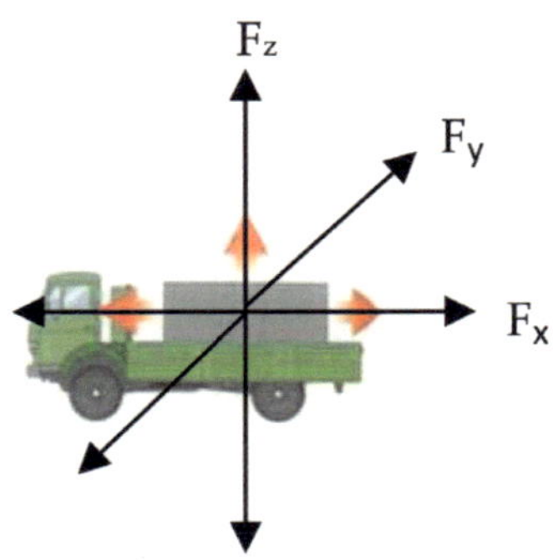

Figure 2. Inertia forces in individual axes. Source: modified from Sdruzeni ridicu [7].

transport generally use a rounded value of 10 ms^{-2}. Such accuracy of input data is not relevant in practice, and the advantage of rounding the gravitational acceleration is that the slightly higher value of the gravitational acceleration ensures a slightly higher value of the resulting inertial force. In the context of the above, the requirement for attachment is stricter, and from the point of view of the liability for damage, those responsible for the loading are protected by a hypothetical "reserve" given by the "stricter" load-bearing requirement.

After verification of the correct cargo attachment, inertia forces at all three axes (x, y, z) are identified. Their values are determined by their relation to the attachment method. For the purposes of this chapter, attachment methods using fastening straps will be further discussed, which are among the most used methods of fixing in road transport, especially for standardized pallet units.

For determining the value of inertia forces, the appropriate formulas (e.g., using the EN 12195–1) serve that best consider the abovementioned facts, i.e., attachment method and attachment means used. The use of fastening straps—the most commonly used attachment method—is further examined, the method of gripping the handling and transport unit (pallet unit). In such cases, EN 12195-1 identifies the formula for calculating only inertia forces in the x-axis (F_x), longitudinal force actuated by the load, and in the y-axis (F_y), transverse force actuated by the load [9]:

$$F_x = \frac{(c_x\text{-}\mu \cdot c_z) \cdot m \cdot g}{2n \cdot \mu \cdot \sin\alpha} \cdot f_s[N] \tag{3}$$

$$F_y = \frac{(c_y\text{-}\mu \cdot c_z) \cdot m \cdot g}{2n \cdot \mu \cdot \sin\alpha} \cdot f_s[N] \tag{4}$$

where c_x, c_y, and c_z are acceleration coefficients in individual axes, μ is the friction factor, m is the mass of the load, g is the gravitational acceleration, f_s represents the safety factor and the number of fastening straps, and finally, the angle α which holds the strap to the floor (horizontal plane).

This calculation can be performed using normative values of acceleration coefficients and friction factors. The accuracy and usability of the normatively determined values of the acceleration coefficients are discussed within the chapter. In the case of friction factor, some differences (inaccuracies) caused by the use of other materials may occur. It is most advantageous to use the friction factor directly from the manufacturer. This is commonly stated, especially in the case of nonslip coating surfaces.

The value of inertia force in z-axis (F_z)—vertical force actuated by the load—is not considered because it is sufficiently assumed to be less than at least one of the above inertia (F_x or F_y). Consequently, the fulfillment of the condition is assumed:

$$(F_z \le F_x) \vee (F_z \le F_y) \tag{5}$$

It is clear from formulas (3) and (4) that the standard is working with a certain degree of abstraction, which is not only due to the absence of calculation of the value of inertia force in z-axis (F_z) but also due to other assumptions. For these reasons, the authors included the safety factor (f_s), which "artificially" increases the value of the resulting inertia forces (for the x- or y-axes) by 10 and 25%, respectively. The following text assumes $f_s = 1.25$. The inclusion of a safety factor in the calculation results in the offsetting of the above abstraction rate and ensures the practical use of the calculation, while it also ensures its relative simplicity.

In practice, software products are used for simplicity, which usually speeds up the determination of inertia forces. However, the basis of calculation and input data is identical to the above-described "manual" approach. Thus software products will no longer be considered, and emphasis will be centered on input data (accelerator coefficient values) and calculation methodology.

The chapter will also focus on input values of acceleration coefficients. Normally, set values of these coefficients are characterized by several shortcomings. The first is their "averaging," i.e., the empirically determined values are statistically evaluated, and the exact method of their determination is not given in EN 12195-1. It can only be assumed that the resulting value of individual acceleration coefficients was based on the mean values that were increased by the "safety factor" analogue, which should include a high rate of abstraction in determining values and, above all, allow their practical use.

In particular, the values of acceleration coefficients do not consider the extreme fluctuations of the individual acceleration coefficients in the transport. To rely on a hypothetical basal vector (the highest empirically mentioned values of the acceleration coefficients in all three axes) would certainly not be effective. However, it should be considered under greater observation. The measurements made from empirical analyses (see below in Chapters 4 and 5) show that, in some cases, especially in roads in poor or bad technical conditions, extreme fluctuations pose a big problem.

Even double and higher rates of normatively determined values of acceleration coefficients are not an exception, and the data found that they represent more than 0.67% of all values, even in the case of the highway. The situation on lower class roads can be expected to be even worse, and the occurrence of extreme fluctuations is more likely.

Another lack of normative values of acceleration coefficients, which to a certain extent is related to the previous, is the non-consideration of infrastructure type (quality). This can be demonstrated (ad absurdum) to compare inertia forces during transport by road (or first-class roads) and off-road transportation (see **Figure 3**).

At the first sight, there is a visible difference between the central part of the chart, which represents the ride on off-roads, and the marginal ones that represent the highway from Brno to Vyškov and, after field training, back from Vyškov to Brno.

Off-road transport has very limited impact upon normal commercial transport and is only interesting for integrated rescue systems or the army. However, based on the data (see below), one may assume a graduation of the magnitude of the acceleration coefficients in relation to the roads of given classes, especially on their quality. The eventual inclusion of off-road conditions can only serve as a logical addition to extremes, identifying both the best considered roads (highways) and the worst alternative (off-road).

Specific road quality of a given class, of course, must be generalized, but the output would offer a set of acceleration coefficients for a given type of infrastructure (road). Such an approach would ensure an effective consolidation concept, i.e., the use of fastening methods

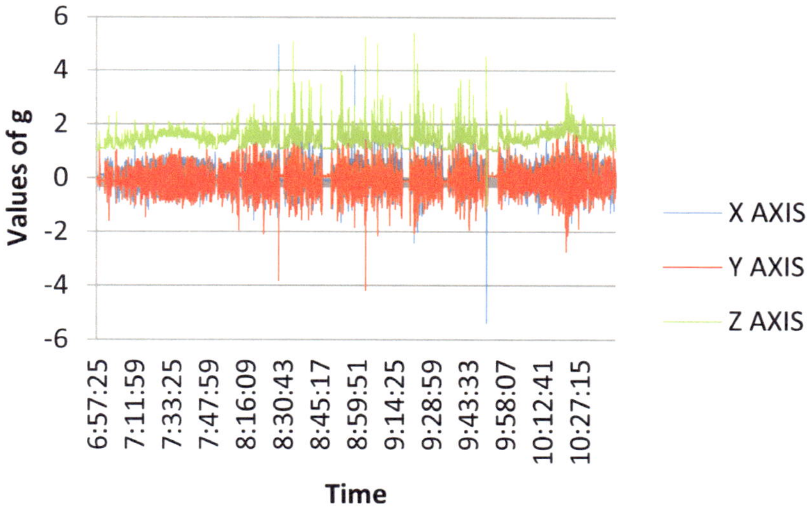

Figure 3. Values of acceleration coefficients (Brno–Vyškov and return).

with such load-bearing capacity of the fastener that it would provide sufficient protection of cargo carried against the inertial forces in all three axes. On the other hand, there would be no second extreme, and it is an "over-dimensioning" fixation according to the philosophy of Just in Case.

Just in Case's philosophy is known primarily from the history of inventory theory [10], i.e., maintaining inventory in every case represents the opposite of Just in Time philosophy. This approach can easily be demonstrated in the case of the use of fastening straps, where the load-carrying capacity of the entire fastening system (more straps) can be measured in a variable manner by a suitable combination of the binding capacity of individual straps and their number. The result would be a use of the optimal number of fastening straps to provide optimal protection against inertial forces in handling and transporting the cargo.

This solution offers the optimal proportion between sufficient fastening to avoid damage to cargo or using other technical means and the number of fastening straps, not only their number but also the time required for their placement and tension, including subsequent removal and reverse logistics.

3. Highways in the Czech Republic and accident rates

In the Czech Republic, there are a total of 55,737.5 km of roads and highways (year 2015). This number has remained virtually unchanged since 2010 (55,751.9 km). Highways represent a relatively small share (only 1.4% from a total length of 776.0 km). Roads of "very good" quality also include the Class I roads, which include expressways. Their total length in 2015 was 6244.9 km. The percentage of highways and Class I roads in the total road network is nearly 12.6%.

However, this category does not include local roads owned by individual municipalities, which are often in a worse technical condition. The total length of local roads is 74,919.0 km. If we compare the length of highways and Class I roads to total length of roads and local roads (130,656.5 km), their share would be very small: less than 0.6% of highways and 5.4% of combined highways and Class I roads [1].

Due to the assumed lower occurrence of extreme variations (values) of acceleration coefficients on highways and Class I roads, the normative values given in EN 12195-1 can be used for these purposes. However, in the remaining 98.6%, respectively, 87.4% of cases may vary and the assumed values of the coefficients of acceleration (or resulting inertial forces) may be higher. Therefore, there is a real risk that cargo will not be appropriately (sufficiently) attached to the vehicle for these purposes. If we include local communications, the situation would dramatically worsen.

Compared to the European Union, the Czech Republic is considered relatively average in road network length, including its length of highways (see Transport Yearbook 2015). The comparison of the Czech Republic with neighboring states is interesting, as some disproportions can be identified here; for example, Poland is significantly larger, both in terms of size and number of inhabitants, yet has only roughly twice the length of highways compared to the Czech Republic. Germany, on the other hand, has a 17 times longer length of highways than the Czech Republic. Austria has more than double and Slovakia only half the length of highways in the Czech Republic.

Over the past 8 years (data available only until 2013), there is a different trend in the volume of road freight transport in the European Union. In the Czech Republic, the increase in tonne-kilometers transported exceeds 26.3% (an increase from 43,447 tkm to 54,894 tkm from 2005 to 2013). For neighboring countries and certain other countries of the European Union, the situation is often different for the same period examined. Poland recorded one of the largest increases in the volume of road freight transport in the European Union, by more than 121.4%. Germany is relatively stagnant, with a slight decrease by 1.4% over the period under examination. Austria recorded a larger decrease of more than 34.6%, and finally, the situation is similar in Slovakia to that in the Czech Republic, and the Slovak volume of road freight transport grew by almost 33.6% [1].

Accidents in road transport are a large long-term problem. In the Czech Republic, despite several preventive measures, the number of accidents failed to fall. Such measures include a suitable setup of the transport system, better information for drivers, as well as measures to reduce the risk of injury or death by road users—especially of a technical nature for vehicles and transport infrastructure.

The total number of road traffic accidents in the Czech Republic in 2005 was 25,239. In 2010, it was only 19,676; however, in 2015, it rose again to 21,561 [1, 11–13].

This increase in the number of road traffic accidents is caused by the steady increase of vehicles (see **Table 1**) registered in the Czech Republic, a comparable situation as in the rest of Europe.

It is clear from **Table 1** that, although there was a slight increase in the number of accidents between 2010 and 2015 (less than 9.6%), the increase in the number of vehicles registered in the

http://dx.doi.org/10.5772/intechopen.70829

Czech Republic has become much more significant in both most important segments (for passenger cars and trucks). This increase was observed between 2010 and 2015, when the number of accidents increased by almost 13.8% for passenger cars and nearly 10.6% for trucks. Compared with 2000 (2000–2015), the increase is very significant: almost 49% for passenger cars and nearly 138% for trucks.

A positive trend is in the number of people killed within 30 days of the accident, which has steadily declined since 2000 (with few exceptions in 2002, 2007, and 2014). In 2015, it dropped to half its level of 2000 (see **Table 2**).

In the case of the number of injured, there was also a significant decrease between 2000 and 2015 (almost 16.9%), yet the trend has increased in recent years. Between 2010 and 2015, this statistic grew by nearly 10.6%.

A certain lack of statistics at the national level is caused by the absence of a distinction between technical causes of accidents and specific types of technical defects. Section 6.2.1 of each transport yearbook classifies road traffic accidents according to location and type, but the technical cause is not shown. These statistics are recorded by certain traffic haulers in their information systems. Unlike aggregated statistical data from the Czech Republic, it is generally not publicly available and thus serves only to manage the operation and vehicle fleet of the relevant entity (hauler) [14].

In other words, if the carrier fails to publicize its own data collection, it is not possible to determine from official statistics how many times a road traffic accident occurred due to improperly or insufficiently secured cargo.

Number/year	2000	2005	2010	2015
Road traffic accidents	25,445	25,239	19,676	21,561
Passenger cars*	3,438,870	3,958,708	4,496,232	5,115,316
Trucks*	275,617	415,101	584,921	646,792

*Vehicles registered in the Czech Republic.
Source: Refs. [1, 11–13].

Table 1. The development of road traffic accidents in the Czech Republic and the number of vehicles in the selected segment.

Number/year	2000	2005	2010	2015
Killed*	1486	1286	802	738
Injured	32,439	32,211	24,384	26,966

*Number of those killed in road traffic accidents within 30 days from the date of the accident.
Source: Refs. [1, 11–13].

Table 2. Number of people killed or injured in road traffic accidents in the Czech Republic.

4. Statistics: evaluation of acceleration coefficients

The basic approach to the statistical evaluation of measured data is a demonstration of extremes—values of acceleration coefficients on the highway and their comparison with values measured on unpaved roads (off-road). Comparison with the normative values of the acceleration coefficients is also part of the comparison.

The basis for statistical evaluation is the graph in **Figure 3**, which illustrates the situation where a group of drivers is transported to a training location on an unpaved road (off-road) and then returns. The chart in **Figure 3** contains raw data, which was subsequently cleaned up by a part of the transport at the city and breaks between the beginning of training rides on unpaved roads (off-road) and between drivers who took part in those trainings. In total, there were six drivers, and one of them provided transport to training from Brno to Vyškov and return.

The aim of the evaluation is a statistical analysis of the measured acceleration coefficients, to find point and interval estimates of the parameters of the partial files and to compare the parameter estimates using statistical tests. The acceleration coefficients for individual axes from the set of measurements taken on the highway and unpaved road (off-road) are compared.

Statistical analysis was performed by the test of normality, which was graphically verified using a Q-Q plot [15]. Normality testing was performed using predetermined skewness and kurtosis coefficients [16]. These tests showed slight deviations from normality, especially during the testing of skewness of distribution; however, the graphic analysis failed to reveal significant deviations from normality. Theoretical quantiles and the corresponding empirical quantiles were approximately in a straight line. For illustration, **Figures 4** and **5** show the Q-Q plots for the given training rides on the unpaved road (off-road) and the measured acceleration coefficients in the x- and y-axes [17].

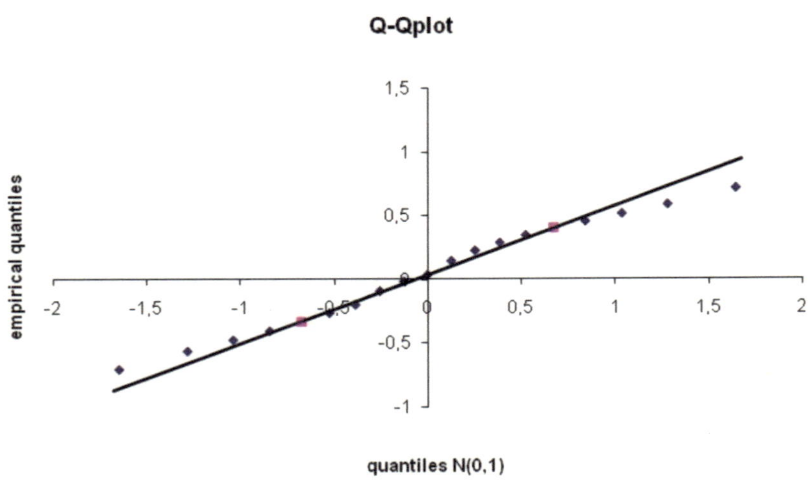

Figure 4. Q-Q plot: unpaved road (off-road)—acceleration coefficients at coordinate x. Source: Vlkovsky et al. [17].

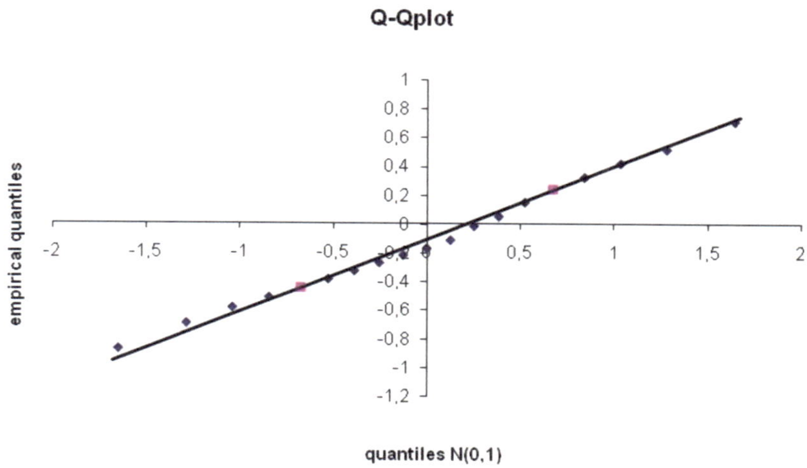

Figure 5. Q-Q plot: unpaved road (off-road)—acceleration coefficients at coordinate y. Source: Vlkovsky et al. [17].

Due to the large scale of compiled sets (more than 5000 data in each set) asymptotic confidence intervals are used, and asymptotic statistical tests are based upon assumptions of asymptotic normality of the observed characteristics [15, 16].

Data is the averaged output of the four measuring devices (accelerometers), and the values are compared with the basal variation of the acceleration coefficients given by the standard [9]:

$$v_n = (0.8, 0.6, 1.0). \tag{6}$$

The measurement for the z-axis started at 1 g, i.e., the value of 1 g of the basal variant corresponds to 2 g on the measuring device. From the measured data, two statistical files were created, which were cleaned by a ride through the city and the vehicle's stop time due to the replacement of the drivers.

The first set was measured on the highway (index 1), and its range is $n_1 = 6148$. The second set was measured on unpaved communication (index 2), and its range is $n_2 = 5257$. For both sets, the basic characteristics of the mean and sample standard deviation were then determined as the expected value μ and standard deviation σ. In addition, for individual axes the number of values was set that exceeded the value given by the standard. The relative frequency of numbers was an estimate of parameter θ, which indicated the probability of exceeding norms. Estimated characteristics matched with both the first set of data obtained by measuring on the highway by index 1 and for the second set of data obtained by measurement on unpaved road (off-road) by index 2. The estimates were supplemented with 95% confidence intervals for each parameter [17].

The comparison of both data files (file 1 and file 2) was performed by comparing the standard deviations σ_1 and σ_2, then comparing the mean values μ_1 and μ_2, and comparing the probabilities θ_1 and θ_2. Comparisons are made for each axis (x, y, z).

Confidence interval for	Acceleration measurement in the axes								
	x			y			z		
	PE	LB	UB	PE	LB	UB	PE	LB	UB
σ_1/σ_2	0.904*	0.881	0.928	0.914*	0.891	0.938	0.778*	0.758	0.799
μ_1-μ_2	−0.008	−0.0251	0.0097	−0.013	−0.0305	0.0050	0.008	−0.0028	0.0188
θ_1-θ_2	−0.0141*	−0.0228	−0.0054	- 0.0157*	−0.0307	−0.0007	−0.0156*	−0.0231	−0.0081

*Means significant difference between parameters of the first and the second dataset at a 5% significance level.
PE—parameter estimation; LB—lower boundary of 95% confidence interval; and UB—upper boundary of 95% confidence interval.
Source: Vlkovsky et al. [17].

Table 3. Statistical tests of equality.

Using these confidence intervals, statistical tests were then performed at the significance level of 5% (significance level) to compare individual pairs of the parameters. Calculated confidential intervals were based on the methodology described in [15, 16].

It can be seen from **Table 3** that the mean values of the acceleration coefficients in each of the axes (x, y, z) between sets are obtained on the highway and unpaved road (off-road). Unpaved roads (off-road) are statistically significant at 5% of the significance level. On the other hand, the variability of the comparison sets is statistically significant in each axis. The variability of acceleration coefficient values measured on the highway is significantly statistically lower than the variability measured on unpaved roads (off-road). Also, the probability values of exceeding the norm in each of the axes are statistically significant. Probability of exceedance is higher (as expected) on unpaved roads (off-road) [17].

5. Transport model

The transport model is designed to illustrate the differences between the evaluated files (see chapter 4) and to demonstrate the effect of the extreme values of the acceleration coefficients on the resulting inertial forces ($F_{x,y}$) that affect the cargo during transport.

The basis of the designed model is the creation of basal variants for both measured sets—the values of the acceleration coefficients detected on the highway and on unpaved roads (in off-road). Basal variants are created as the average of the three largest (in absolute value) fluctuations in each of the axes within a given set. Input values are summarized in **Table 4**, including the time at which the fluctuations occurred. Values in z-axis are cleared by 1 g in terms of displacement of the measurement axis, i.e., the value of gravitational acceleration (1 g).

Based on the highest values in **Table 4** and their average, basal variations are created for both data files (n_1, n_2). Shock direction—the sign is not considered; the value of the acceleration

	Acceleration measurement in the axes-extremes								
	x			y			z		
Time(n_1)	10:22:07	10:22:44	10:23:01	10:22:08	10:22:12	10:23:01	10:22:07	10:22:08	10:22:44
Value (n_1)	−2.0900	−1.7225	−1.9075	−2.7425	−2.3675	2.2650	2.3725	2.5650	2.1625
Time(n_2)	8:29:14	8:59:07	9:51:01	8:29:14	8:35:01	9:03:27	8:35:01	9:03:27	9:22:36
Value (n_2)	4.9675	4.2175	−5.4000	−3.8200	2.3500	−4.1825	4.1000	4.2700	4.4225

Table 4. Input values of basal variants.

coefficient is used, i.e., all values in **Table 4** are taken at absolute value. The resulting basal variants are:

$$v_1 = (1.9067, 2.4583, 2.3667). \tag{7}$$

$$v_2 = (4.8617, 3.4508, 4.2642). \tag{8}$$

The model is created under the following assumptions:

- the cargo is transported (attached) to a Tatra T-810;
- the cargo consists of a model pallet unit, which has a height of 1600 mm and is designed using a standard EUR pallet 1200 × 800 mm;
- the cargo on the pallet unit is fixed by default use of shrink film, and the fastening method is not the subject of the model;
- cargo weight of the pallet unit is 500 kg;
- the pallet unit is fastened with commonly available tie-down straps; the subject of the model does not determine the number or type of tie straps;
- the pallet unit is located longitudinally in relation to the direction of movement of the vehicle (x-axis);
- the source of data is created by basal variants (see above), and the normative values of the acceleration coefficients are given by EN 12195-1; and
- the formula and other necessary input data from EN 12195–1 are used for the calculation.

The model pallet unit and the method of its attachment are shown in **Figure 6**, from which it is possible to determine the angle α required to determine the magnitude of inertia forces.

From **Figure 6**, it is possible to determine the angle α from the known width of the vehicle's cargo space (2506 mm) and parameters of pallet unit (1200 × 800 × 1600 mm). The distance of the pallet unit heel to the anchor point (k) is then:

$$(2506 - 800)/2 = 853 \text{ mm}. \tag{9}$$

Using the trigonometry function, tangent can be calculated angle α with the height of pallet unit (v) and the distance of pallet heel from anchoring point (k):

$$tg\alpha = v/k \tag{10}$$

after using values:

$$tg\alpha = 1600/853 \tag{11}$$

$$\alpha = 61.94^{\circ} \tag{12}$$

The input values of the model for fitting into formulas (3) and (4) are shown in **Table 5**. For clarity, the magnitude of the inertia forces in the x- and y-axes will be examined. The F_z value is

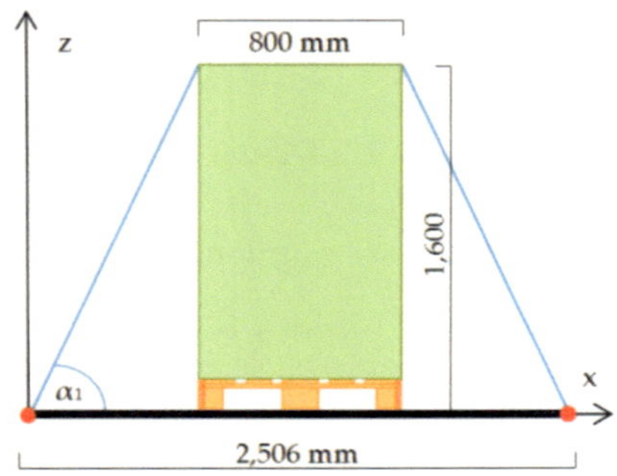

Figure 6. Method of fixing the model pallet unit. Source: Vlkovsky et al. [18].

Variable	Value			Unit	Note
	Norm	Highway	Off-road		
$F_{nx,y}$	?	—	—	N	Inertial force – norm
$F_{1x,y}$	—	?	—	N	Inertial force – highway
$F_{2x,y}$	—	—	?	N	Inertial force – off-road
c_x	0.8000	1.9067	4.8617	—	Coefficient of acceleration (x)
c_y	0.6000	2.4583	3.4508	—	Coefficient of acceleration (y)
c_z	1.0000	2.3667	4.2642	—	Coefficient of acceleration (z)
μ	0.4	0.4	0.4	—	Coefficient of friction
m	500	500	500	kg	Mass of cargo
g	9.81	9.81	9.81	ms^{-2}	Gravitational acceleration
f_s	1.25	1.25	1.25	—	Safety factor
n	1	1	1	pc	Number of lashing straps
α	61.94	61.94	61.94	°	Angel – among strap and floor

Table 5. Input values of transport model.

http://dx.doi.org/10.5772/intechopen.70829

not calculated because formula (5) is assumed to be valid. The value of the inertia forces in the x- and y-axes will be calculated using the normalized acceleration coefficients (v_n), basal variant 1 (v_1), and basal variant 2 (v_2).

The following designation is used to differentiate individual inertia forces with different input data:

- F_{nx}—inertial force in the x-axis using normatively determined acceleration coefficients (v_n)
- F_{ny}—inertial force in the y-axis using normatively determined acceleration coefficients (v_n)
- F_{1x}—inertial force in the x-axis using basal variant 1 (v_1)
- F_{1y}—inertial force in the y-axis using basal variant 1 (v_1)
- F_{2x}—inertial force in the x-axis using basal variant 2 (v_2)
- F_{2y}—inertial force in the x-axis using basal variant 2 (v_2)

After using the three variants of input data into formulas (3) and (4), the results are shown for overview at **Table 6**.

As apparent from the results in **Table 6**, acceleration coefficients significantly affect the resulting inertia force influencing cargo (affecting the fastening strap within the transport model). When using acceleration coefficients stipulated as normative, the cargo "behaves" according to expectations, meaning that resulting inertia forces in the x- and y-axes are relatively small. Even with the inclusion of the safety factor, they do not correspond to as little as 5000 N, which basically corresponds to the cargo weight (500 kg).

Forces, ratio	Value	Unit	Note
F_{nx} (for v_n)	3.474	N	—
F_{ny} (for v_n)	1.737	N	—
F_{1x} (for v_1)	8.338	N	—
F_{1y} (for v_1)	13.128	N	—
F_{2x} (for v_2)	27.410	N	—
F_{2y} (for v_2)	15.156	N	—
$F_{1x} : F_{nx}$	2.40	—	Inertial ratio in the x-axis – norm:highway
$F_{2x} : F_{nx}$	7.89	—	Inertial ratio in the x-axis – norm:off-road
$F_{2x} : F_{1x}$	3.29	—	Inertial ratio in the x-axis – highway:off-road
$F_{1y} : F_{ny}$	7.56	—	Inertial ratio in the y-axis – norm:highway
$F_{2y} : F_{ny}$	8.73	—	Inertial ratio in the y-axis – norm:off-road
$F_{2y} : F_{1y}$	1.15	—	Inertial ratio in the y-axis – highway:off-road

Table 6. Values of inertial forces in x and y axes and their ratios.

Forces	Value	Unit	Note
F_{nx}	3.474	N	For normative values of c_x, c_y, and c_z
F_{ny}	1.737	N	For normative values of c_x, c_y, and c_z
F_{1x}	−9.215	N	Measured at 10:22:08, for $c_x = -0.0350$, $c_y = -2.7425$, and $c_z = 2.5650$
F_{1y}	−32.729	N	Measured at 10:22:08, for $c_x = -0.0350$, $c_y = -2.7425$, and $c_z = 2.5650$
F_{2x}	−43.051	N	Measured at 9:51:01, for $c_x = -5.4000$, $c_y = -0.5925$, and $c_z = -1.1075$
F_{2y}	−1.298	N	Measured at 9:51:01, for $c_x = -5.4000$, $c_y = -0.5925$, and $c_z = -1.1075$

Table 7. Values of inertial forces – extremes in x and y axes.

However, when hypothetic basal variants v_1 and v_2 are used, the situation becomes very different. The magnitude of the inertia forces is significantly greater on both axes, both in the case of highway and unpaved road (off-road).

When considering the greater of the affecting inertia forces, the required securing force of the strap is in the case of a highway 7.56× greater (F_{1y}) than in the case of an inertia force arising from normative-stipulated values of acceleration coefficients (F_{ny}). On an unpaved road (off-road), the situation is presumably even worse because an inertial force value of 27,410 N (F_{2x}) on the x-axis was calculated from the measured values, corresponding to nearly eight times the value of the inertial force calculated based on norm-stipulated values of acceleration coefficients (F_{nx}). In a simplified way, one might say that, using this extreme value, a cargo "behaves" as if it weighed 2741 kg instead of 500 kg.

Should the data from the moment of transportation (including sustaining the direction of impact) be used to construct the model, where the greatest acceleration coefficient value was measured (−2.7425 in the case of a highway and −5.4000 for off-road conditions, respectively), the situation would be as follows (see **Table 7**).

As apparent from **Table 7**, extreme inertial forces affect the cargo in isolated cases, fundamentally exceeding expected values. In the case of a value of −5.4000 for off-road conditions, the value of the respective inertial force affecting the cargo (securing strap) is −43.051 N. If we abstract it from its direction, it is 12.39× greater than the expected value corresponding to the acceleration coefficients from the EN 12195–1 norm. The cargo therefore "behaves" as if it weighed 4305 kg.

6. Conclusion

The importance of researching cargo securing in road transportation is apparent both from the study presented (chapter) and, in particular, from the estimated number of accidents caused by improperly or insufficiently secured cargo. The significance of cargo securing also rises in the context of multimodal transportation where the manipulation and transportation units must be prepared for the transportation effects in more than one form of transportation. With few exceptions, road transport is a necessary part of the door-to-door concept of multimodal transportation.

The initial requirement should be the collection of relevant data concerning accidents that are typically unavailable on national level. A specific overview regarding the number of accidents caused by improper or insufficient cargo securing would enable a strict focus on the specific causes of accidents and taking adequate measures. Such statistics would also verify, or disprove, the significance of impact (acceleration coefficients) on various types of surfaces, in different countries and for various types of vehicles and cargo.

However, the problem is not only in obtaining the data but also in their suitable statistical processing. Each isolated differentiation should not be considered in the case of a common cargo, but statistically significant divergences should be included in calculations. Further analysis including occasional extreme divergences is necessary in the case of sensitive cargo (dangerous, fragile, etc.) that may cause extensive damage to property and injuries to the driver or other road traffic participants.

A separate area for further study is also the services of the integrated emergency system, as well as the army, where the requirements are very specific. The outputs, particularly for unpaved roads (off-road conditions), would be particularly effective and useful for these two segments.

Another area of study will be the differentiation of approaches to cargo securing according to the expected transport route—road networks used. This is an analogy to the multimodal transportation where, at least implicitly, the "worst" form of transportation is considered. In this case, from the perspective of the basal variant, the form of transportation where the effect of inertial forces is expected to be greatest is considered "worst." The resulting model would reflect magnitudes of acceleration coefficients for various road networks, for example, according to standard classification:

a. Highways, speed roads, and Class I roads

b. Class II and III roads (possibly including local roads)

c. Unpaved roads (off-road conditions)

The respective basal variant would be used for the category given, having been identified using sufficient scientific methods. Aside from standard statistics tools, there is also an option to use the method of spectral analysis or possibly Fourier transform.

The use of upgraded software supposes that it would serve as a welcome addition to existing software and could reflect the type of road, specifically the variants mentioned, and would be of indisputable advantage.

Software support should enable the verification of expected inertia forces, as well as dynamic simulation, for example, on the basis of the MSC ADAMS Multibody Dynamics (MBD) software that provides the opportunity to create mathematical models of transport vehicles and analyze cargo stress during vehicle movement.

Acknowledgements

The chapter was written with the support of long-term strategy of organization development ROZVOLOG, funded by the Ministry of Defence of the Czech Republic.

Author details

Martin Vlkovský

Address all correspondence to: martin.vlkovsky@unob.cz

Department of Logistics, University of Defence, Brno, Czech Republic

References

[1] Ministry of Transport of the Czech Republic. Transport Yearbook [Internet]. 2015. Available from: https://www.sydos.cz/cs/rocenka-2015/yearbook/htm_uk/index.html [Accessed: Jul 12, 2017]

[2] Fenollar Solvay A, Haberstroh M, Thelen S, Schilberg D, Jeschke S. New intermodal loading units in the European transport market. In: Jeschke S, Isenhardt I, Hees F, Henning K, editors. Automation, Communication and Cybernetics in Science and Engineering 2015/2016. Springer: Cham; 2016. p. 687-697. DOI: 10.1007/978-3-319-42620-4_52

[3] CSPSD. Mobile Expert Units. [Internet]. 2017. Available from: http://www.cspsd.cz/mobilni-expertni-jednotky [Accessed: Jul 15, 2017]

[4] European Commission. Road Safety: Best Practice Guidelines on Cargo Securing and Abnormal Transport. Luxembourg: Publications Office of the European Union; 2014. p. 96 DOI: 10.2832/69096

[5] ČSN ISO 668. Series 1 Freight Containers – Classification, Dimensions and Ratings. Prague: Czech Office for Standards, Metrology and Testing; 2015. p. 20

[6] AMovP-5. Multimodal Transport Issues. Capallen: North Atlantic Treaty Organization; 2005. p. 148

[7] Sdruzeni ridicu CZ. Cargo Securing – Basic information. [Internet]. 2017. Available from: http://soubory.proridice.eu/naklady/upevneni_nakladuCZ.pdf [Accessed: Jul 16, 2017]

[8] Wolfram Alfa Widgets. Gravitational Acceleration. [Internet]. 2017. Available from: http://www.wolframalpha.com/widgets/view.jsp?id=d34e8683df527e3555153d979bcda9cf [Accessed: Jul 15, 2017]

[9] ČSN EN 12195-1. Load Restraining on Road Vehicles – Safety – Part 1: Calculation of Securing Forces. Prague: Czech Office for Standards, Metrology and Testing; 2011 p. 48

[10] Army Logistician. Eliminating the Iron Mountain. [Internet]. 2017. Available from: http://www.alu.army.mil/alog/issues/JulAug04/C_iron.html [Accessed: Jul 14, 2017]

[11] Ministry of Transport of the Czech Republic. Transport Yearbook [Internet]. 2010. Available from: https://www.sydos.cz/cs/rocenka-2010/yearbook/htm_uk/index.html [Accessed: Jul 12, 2017]

[12] Ministry of Transport of the Czech Republic. Transport Yearbook [Internet]. 2005. Available from: https://www.sydos.cz/cs/rocenka-2005/yearbook/htm_uk/index.html [Accessed: 2017–07-12]

[13] Ministry of Transport of the Czech Republic. Transport Yearbook [Internet]. 2000. Available from: https://www.sydos.cz/cs/rocenka-2000/yearbook/html/start_e.htm [Accessed: 2017–07-12]

[14] Vlkovsky M, Vlachova H. Cargo securing and accident rate in the army of the Czech Republic. In: The 23rd International Conference The Knowledge-Based Organization: Applied Technical Sciences and Advanced Applied Technical Sciences and Advanced Military Technologies (KBO 2017); 15–17 June 2017; Romania. Sibiu: KBO; 2017. p. 148-154

[15] Johnson RA, Wichern DW. Applied Multivariate Statistical Analysis. New Jersey: Prentice-Hall; 2007. p. 773

[16] Neubauer J, Sedlacik M, Kriz O. Principles of Statistics: Applications in Technical and Economic Disciplines. 2nd extended ed. Prague: Grada; 2016. p. 278

[17] Vlkovsky M, Smerek M, Michalek J. Cargo securing during transport depending on the type of road. In: World Multidisciplinary Civil Engineering-Architecture-Urban Planning Symposium – Procedia Engineering (WMCAUS 2017). 12–16 June. Czech Republic. Prague: WMCAUS; 2017. p. 2017

[18] Vlkovsky M, Rak L, Taksas B. Cargo securing during transportation – Using extreme values. In: The 23rd International Conference The Knowledge-Based Organization: Applied Technical Sciences and Advanced Applied Technical Sciences and Advanced Military Technologies (KBO 2017); 15–17 June 2017; Romania. Sibiu: KBO; 2017. p. 142-147

Chapter 4

Joint Road Safety Analysis in Open Roads and Tunnels

Carlo Polidori, Adewole Adesiyun,
Antonio Avenoso, Kallistratos Dionelis, Liljana Cela,
Christophe Nicodème and Thierry Goger

Additional information is available at the end of the chapter

http://dx.doi.org/10.5772/intechopen.71734

Abstract

From the user's (drivers) point of view, a road is a unique linear infrastructure although it is clear for a driver that such a linear infrastructure consists of open terrain (open roads) and occasionally closed environment (tunnels). Since the two environments present different safety issues, the related analyses are usually conducted on the field by different experts; those conclusions are sometimes not well interlinked and harmonised, leading to safety gaps particularly, but not only, in the transition areas. Joint safety analyses conducted in the same time by a group of road safety and tunnel safety experts can fill such a gap and increase the safety level of the whole infrastructure. During the year 2016, an international group of road safety experts and tunnel experts visited five road sections with open roads and tunnel in Europe and performed joint safety analyses together with the infrastructure managers. Such analyses were conducted according to a predefined experimental procedure to check the effectiveness of the joint analyses with respect to the usual ones. The key results are that joint safety operations in tunnels and open roads are possible and extremely useful: their cost can be very low when well planned.

Keywords: road safety inspection, open roads, tunnels, transition areas, joint safety operations

1. Introduction

The issue of road safety inspections in tunnel was discussed in two workshops held at the European Social and Economic and Social Committee (EESC) by a group of international stakeholders in February and May 2013: a debate that was initiated about operations such as Road Safety Audit (RSA) during the design process or Road Safety Inspection (RSI) after opening to traffic, according to the prescriptions of the Directive 2008/96/EC (Road Infrastructure Safety Management – RISM). Such operations could be beneficial for risk prevention in tunnels, but the Directive 2008/96/EC does not apply to road tunnels, which are covered by Directive 2004/54/EC (on tunnels safety management): a formal interpretation of the two Directives may lead to a barrier to the safety inspection of the whole infrastructure.

In fact, from the user's (drivers) point of view, a road is a unique linear infrastructure although it is clear for the driver that such a linear infrastructure consists of open terrain (open roads) and occasionally closed environment (tunnels). The driver wants to receive the same high safety levels, without being interested to know all the details of the 'infrastructure safety chain' that produces such safety.

On the other hand, from the infrastructure managers' point of view, the road is surely not a unique linear infrastructure, because producing and managing safety in a closed environment (tunnels) is much more demanding and extreme than in an open road infrastructure.

Due to different characteristics of a tunnel (e.g. level of visibility, design, enforcement of traffic regulations, etc.), it is important to look at the safety perception of drivers in this environment, otherwise called—subjective safety [1] in normal conditions, as well as in critical scenarios. The users' perspective was also analysed in 2010 [2], revealing several causes of fear induced to users.

In order to find a common harmonised approach in tunnel and open road safety management, during the year 2016, an international group of road safety experts and tunnel experts visited five road sections with open roads and tunnel in Europe and performed joint safety analyses together with the infrastructure managers in the framework of the European Project ECOROADS [3] (**Table 1**).

Test site, country	Dates of the joint visits	No. of experts (core team)	No. of observers	No. of other experts	Tunnel type and length	Length of open road inspected
Kennedy Tunnel, Belgium	07–08 March 2016	3	3	6	Two tubes, 690 m each	1200 m
Krrabe Tunnel, Albania	05–06 April 2016	4	4	5	Two tubes, 2230 and 2500 m	1500 m
Tunnel Rennsteig, Germany	17–18 August 2016	3	3	4	Two tubes, 7916 m each	400 m
Tunnel Strazevica, Serbia	27–28 September 2016	3	1	12	Single tube 745 m	650 m
Tunnel Demir Kapija, Former Yugoslav Republic of Macedonia	18–19 October 2016	4	0	9	Single tube 554 m	400 m

Table 1. Basic data of the five joint safety operations.

http://dx.doi.org/10.5772/intechopen.71734

Such analyses were conducted according to a predefined experimental procedure foreseeing the figure of a 'facilitator' in charge to maintain the contacts between the group of experts, several external observers, and the infrastructure manager, as well as to ensure an adequate feedback after the visit (**Figure 1**).

The effectiveness of the joint analyses respect to the usual ones was fully confirmed, and their main aspects related to the definition of the transition area and common check lists are reported in the following sections.

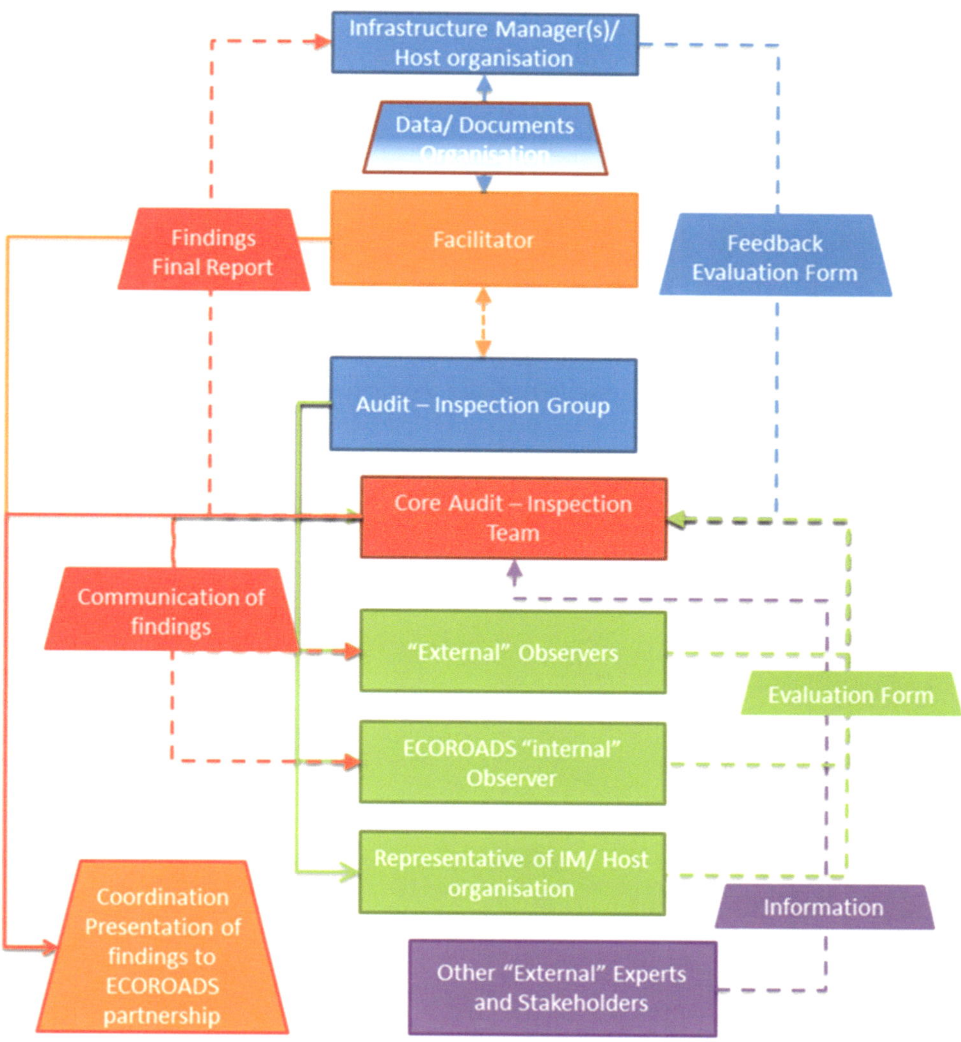

Figure 1. Roles and responsibilities of parties involved in the joint safety operations.

2. Definition of the 'transition area'

For the scope of the joint safety analysis, the transition area between an open road and a tunnel covers areas both inside and outside the tunnel.

As a minimum requirement, it is intended to be the sum of:

a. The distance calculated as the distance covered in 10 s by a vehicle travelling at the speed limit before the tunnel portal, and

b. the stopping distance inside the tunnel after the portal, for a vehicle travelling at the speed limit, if not identical with design speed. Such a distance has been defined in the ECOROADS project by using a longitudinal friction coefficient of 0.40 (wet surface, high speed) and a 2-s reaction time. This gives the following stopping distances from various vehicle speeds (**Table 2**):

This minimum rule obviously applies to the opposite direction and also—maybe slightly modified due to reduced speed within the tunnel—at the exit of the tunnel and in the same direction (**Figure 2** for each direction).

This area, actually its length, may only be extended after a common agreement of the expert team during a joint safety operations' briefing meeting, who provided a sound justification for its modification. This need may emerge from other parameters, such as the speed mentioned above, road marking, signage, lighting, infrastructure design, and should then be considered.

Approaching speed (85th percentile)	**70 km/h**	**90 km/h**	**120 km/h**
Stopping distance (with t = 2 s, fl = 0.40, no grade %)	87 m	130 m	208 m

Table 2. Stopping distances.

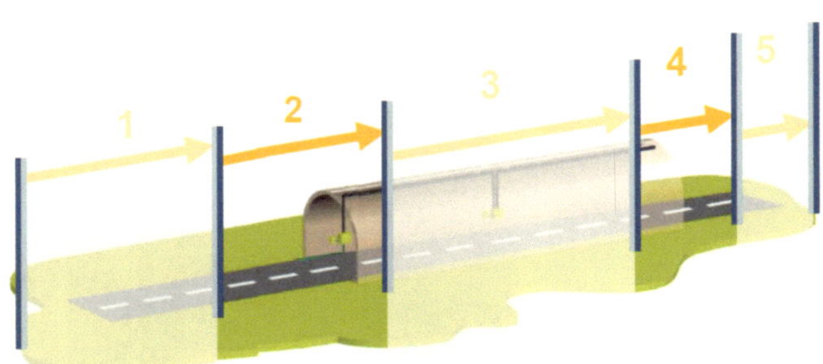

Figure 2. Transition areas are represented by sections 2 and 4.

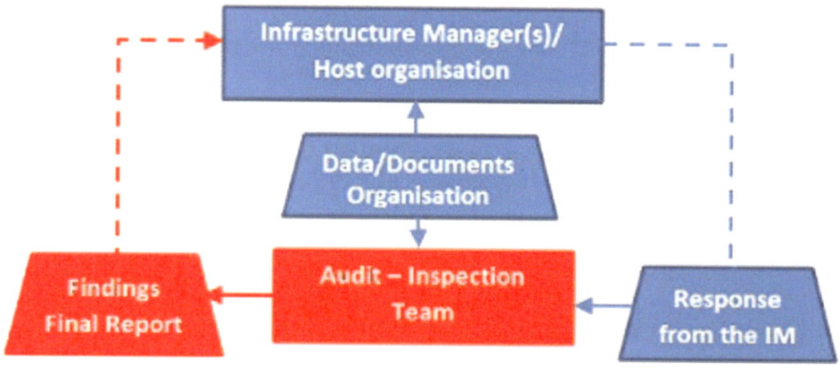

Figure 3. Simplified scheme of the joint safety operations.

Five joint safety operations performed during the ECOROADS project highlighted that such transition areas often presented the more critical road safety issues.

Since the common procedures adopted by the project have been validated through the success of the joint safety operations, the simplified scheme in **Figure 3** can be successfully adopted by each infrastructure manager by using the checklist described in Section 4.

3. Conclusions

When managing the real traffic flows in the real infrastructure, there is a need for coordinated actions. This particularly applies to the transition areas where two different infrastructures ('open roads' and 'tunnels') meet, which leads to the need to develop a harmonised safe traffic management.

During the operations in the five test sites, there was a good level of involvement of infrastructure managers (of tunnels and open roads) and the core group as defined in the scheme of the previous **Figure 1**.

A multitask procedure was adopted to allow an experimental deployment of a multidisciplinary and multifunctional team of international experts.

ECOROADS has been funded by the HORIZON 2020 Programme [5]; the project collected 42 feedback forms from the expert group. Due to the rotation of the figures involved, all the members of the core groups were observers in at least one joint safety operation. There was widespread consensus on the following added value of the joint safety operations:

- Working in a mixed team (safety/tunnel experts), mixture of experiences from different countries

- Common/coordinated approach for open roads—transition area—tunnel in one project that guarantees a harmonised safety approach in the traffic management, respecting the different technical characteristics in each area
- Exchange of knowledge and best practices
- Opportunity to visit and see the tunnel from inside and see traffic and driver's behaviour both inside and outside the tunnel ('feel the traffic on my own')
- Opportunity to examine the tunnel without traffic
- Close collaboration between road safety experts and tunnel safety experts regarding the common view of the transition area as a whole
- Mainly, the view of road safety experts on the part of the road in the tunnel with its specific characteristics can be very conducive to evaluate the total safety of the road in a closed environment (tunnel).

Acknowledgements

The study of the joint safety operations has been financed by the European Research and Innovation programme HORIZON 2020, through the project ECOROADS.

A. Appendix: check list for transition areas

Before and after tunnel entrance/exit, according to **Figure 2**; checklists for open roads can be found in [4], and the ones for tunnels can be found in [3].

Site: ….

Direction: ….

Date: … … ….. Time: … ..

Weather conditions: … … … …..

Traffic conditions: … … ….

Expert's name: … ….

Role (Roads/Tunnels): ….

Transition Area (open road tunnel or vice versa): from km …. + ….…. to km …. + …..….

Please use a separate check list for the opposite direction or in the case of multiple road-tunnel sections (**Table 3**).

http://dx.doi.org/10.5772/intechopen.71734

Characteristics	No.	Questions	Yes (✓) No (X)	Comments
1. Function, operating elements, and surrounding/ road environment	1	Is there any information about previous RSA/ RSI or other safety assessment (Tunnel Safety Inspection) final results **relevant to this area**? Have these results and the relevant recommendations been taken into consideration?		
	2	Are there any issues from accident data if available?		
	3	Is the Tunnel - a twin tube tunnel (two monodirectional tubes), - a single tube tunnel, bidirectional - an urban Tunnel		
	4	Is the design or the current situation of the tunnel according to its category?		
	5	Are there radio rebroadcasting frequencies and is the radio frequency adequately communicated to the users?		
	6	Is there information provided to the users about their correct behaviour while driving through a tunnel (use headlights, avoid sunglasses, keep distances, observe signs and signals, switch the radio on, and tune to indicated frequency)?		
	7	Is there information provided to users about the possible incidents in tunnels and appropriate reaction of the users?		
	8	Are there special lanes for HGVs or reduced speed limit for trucks? Is this affecting traffic flow? Is overtaking of trucks prohibited?		
	9	Is there water supply provided near portals?		
	10	Is there gradual reduction of speed from open road to the tunnel speed limit?		
	11			
	12			
	13			
2. Portal	1	Does the tunnel portal have a safe design for all vehicle types? Funnel shape and gradual height reduction is recommended.		
	2	Is there a risk of heavy vehicles hitting the tunnel ceiling or walls?		
	3	Are the portals sufficiently shielded?		
	4	Does the entrance have a slowing down effect due to its design (not informative, dangerous, confining)?		
	5			
	6			
	7			

Characteristics	No.	Questions	Yes (✓) No (X)	Comments
3. Cross section	1	Is the paved area narrowed and lateral clearance reduced at entrance causing speed reduction and driving in more distance to tunnel sidewalls?		
	2	Is the same number of traffic lanes maintained outside and inside the tunnel?		
	3	Are the different traffic lanes' widths sufficient?		
	4	Is crossing the central reserve ensured outside each portal?		
	5	In bidirectional tunnels, are appropriate means used along the median for separation of the two directions of traffic?		
	6			
	7			
4. Alignment, horizontal, and vertical	1	Is the alignment towards tunnel satisfactory (straight, not on curve)?		
	2	Is the crossfall appropriate?		
	3	Is the longitudinal gradient >3%? Is it ≤5%, according to the Directive? Or geographically this was not possible?		
	4			
	5			
	6			
	7			
	8			
	9			
5. Intersections/ Interchanges	1	Are there junctions inside or before-after the tunnel?		
	2	Is direction signing adequate?		
	3	Are there dangerous weavings?		
	4			
	5			
6. Traffic signals/ITS	1	Are there traffic signals?		
	2	Are signals well ahead the portal (150–200 m), where the drivers do not pay attention, as they are focused to the tunnel entrance?		
	3	Can drivers see the traffic signals ahead of the portals in good time? Are they before the entrances to ensure the closure of the tunnel in case of emergency?		
	4	Are there VMS systems? Which are there categories of information (congestion/breakdown/accident/fire/open-close-divert/speed limit/lanes allocation to vehicle categories)?		
	5	Are the VMS messages understandable? Are they only in text or accompanied by pictograms?		
	6	Is it possible to post speed limit signs at sufficient intervals to safely reduce driving speeds?		
	7	Is there need for ATC?		
	8			
	9			
	10			

http://dx.doi.org/10.5772/intechopen.71734

Characteristics	No.	Questions	Yes (✓) No (X)	Comments
7. Traffic signing, marking, and lighting	1	Is there proper distance between signs, harmonised with the driving speed?		
	2	Are there any objects/reasons of distraction of drivers from concentrating on tunnel entrance (irrelevant signs/advertisements?		
	3	Are signs succinct in form and repeated for clarity?		
	4	Are signs perceivable, simple, readable, credible, reliable, memorable, and easily to understand (also for non-nationals)?		
	5	Are signs and signals well dimensioned, coloured, lighted, and repeated frequently in a proper way?		
	6	Is information system complicated? Are written instructions numerous?		
	7	Are there warning signs before the last interchange that a tunnel is ahead to avoid stressful driving, over-height vehicles, and provide exit of vehicles carrying dangerous loads?		
	8	Is horizontal signing used at the roadside edge?		
	9	Road markings – Are profiled markings used? Is there a need for LED lighting?		
	10	Are rumble strips (acoustic lane markings) located ahead of the portals?		
	11	Is the right illumination level chosen at the entrance zone, transition zone? Does the illumination level satisfy all lanes?		
	12	Is the entrance into the tunnel adequately lighted?		
	13	Is there adaptation lighting (adjustment of the light intensity level) at the beginning and end of the lighting/tunnel?		
	14	Are the sight conditions at entrance poor (ocular blinding, stray light, etc.)?		
	15	Is the vehicles' headlights usage obligatory and this clearly communicated with appropriate signing?		
	16	Is the entrance zone designed to provide as low adaptation luminance as possible?		
	17			
	18			
	19			

Characteristics	No.	Questions	Yes (✓) No (X)	Comments
8. Roadside features and passive safety installations (incl. plantings, civil engineering structures, and other obstacles)	1	Are there restraint systems/ guardrails?		
	2	Is there a need for guardrail in the portal zone? And is the guardrail in case well anchored and extended back both outside and inside the tunnel?		
	3	Are there passive objects for minimising the consequences of accident at the portal area? Are there crash cushions and impact attenuators? Of which type?		
	4	Are there lay-bys for emergency, height controls etc., ahead of the portal?		
	5	Are there emergency phones?		
	6	Are there buildings, installations, and equipment for operational reasons (e.g. winter service)? Are their location (distance from carriageway—more than 7 m aside, distance from portal—not more than 50 m), architectural design, access roads, and other characteristics appropriate?		
	7	Are there paths for service or emergency vehicles to U-turn before the tunnel entrance? Are they grade separated (bidirectional tunnels) or at grade (unidirectional)?		
	8	Is there need for refurbishment to avoid risks for users from degradation of outdated equipment?		
	9			
	10			
	11			
	12			
	13			
	14			

Table 3. Check list.

Author details

Carlo Polidori[6]*, Adewole Adesiyun[1], Antonio Avenoso[2], Kallistratos Dionelis[3], Liljana Cela[4], Christophe Nicodème[5] and Thierry Goger[1]

*Address all correspondence to: c.polidori@libero.it

1 Forum Des Laboratoires Nationaux Europeens De Recherche Routiere, Brussels, Belgium

2 European Transport Safety Council, Brussels, Belgium

3 European Association with Tolled Motorways, Bridges and Tunnels, Brussels, Belgium

4 South East Europe Transport Observatory, Belgrade, Serbia

5 The European Union Road Federation, Brussels, Belgium

6 Italian Association of Road Safety Professionals, Roma, Italy

References

[1] SWOV. Subjective safety in traffic. Fact sheet. Institute for Road Safety Research, the Nether-lands, 2012. 2012. Retrieved from: http://www.swov.nl/rapport/Factsheets/UK/FS_Subjective_safety.pdf

[2] Matzke W. How safe can a tunnel be—How safe will it be?—Users' perspective. A dangerous driving companion. 5th International Conference "Tunnel Safety and Ventilation. Graz; 2010. Retrieved from: http://lampx.tugraz.at/~tunnel2016/history/Tunnel_2010_CD/pdf_Dateien/35_Day.pdf

[3] ECOROADS "Effective and Coordinated Roads Infrastructure Safety Operations" - www.ecoroadsproject.eu

[4] PILOT4SAFETYproject - http://pilot4safety.fehrl.org

[5] European Union HORIZON 2020 Programme for Research and Innovation - https://ec.europa.eu/programmes/horizon2020/

Empowering Added Value in Highway Project: A Strategy to Improve the Feasibility

Mohammed Ali Berawi

Additional information is available at the end of the chapter

http://dx.doi.org/10.5772/intechopen.71778

Abstract

Most mega-project infrastructure such as highway construction requires an enormous amount of cost. This situation might be a problem to developing countries that have limited national budget plan. On the other hand, the capability to transfer the funding of infrastructure depends on the project attractiveness to the private sector. The evaluation to involve in the project should be supported by a significant value for money from the business perspective. Optimum feasibility plays a vital role in bridging the partnership between the government and private interests. The study uses a case study of highway project development of Trans-Sumatera highway road (TSHR) that spans about 2700 km to elaborate on how project feasibility can be improved through creative method. Value engineering (VE) approach as a systematic way to generate innovative ideas by combining multidiscipline backgrounds perspective is used. The study shows that additional functions for the project include integration with motorbike toll road, dry port, and railway line; others also include tourism, fiber optic, and service area development. The innovative ideas have contributed to the significant increase in the internal rate of return (IRR) to the project from 7.79 to 12.76%. The study also formulates institutional scheme through build operate sharing transfer (BOST) which administers government's role in the project development.

Keywords: infrastructure, feasibility, highway, innovation, mega-project, value engineering

1. Introduction

Highway encourages connectivity among regions to expand economic development in a nation. It also eases the traffic flow in developing areas, improves the distribution of goods and services, and provides equity development among regions in a country [1, 2]. The benefits of highway construction directly affect the mobility and accessibility of people and goods and time travel as well as the other benefits of the regional competitiveness and national gross domestic product [3, 4].

Countries such as France, Spain, the United States, Canada, South Africa, China, and Australia have invested in highway project to support their economic enlargement [5]. The construction of infrastructure is argued as one of the solutions to improve the level of competitiveness in price logistics. The strategy will increase more transactional activities and contribute to the macroeconomic development.

Indonesia is a developing country located in South East Asia region, which has a population almost 262 million inhabitants in 2016 [6]. According to the Ref. [7], the state attempts to construct toll road project since 1973 and currently operates 987.44 km that span over the country. Despite its massive construction during the past five years by the government, the total operated highway is left behind from other nations. Japan has developed 8358.3 km of expressway [8]; Korea manages 4193 km of the toll road and expects to operate 6160 km in 2020 [9] and Malaysia as the neighboring country already has 1820 km of toll highways [10]. As the motorway plays a significant role in national competitiveness, expansion of the project is required in Indonesia.

The country attempts to build new highway road in the western part of the country called Trans-Sumatera Toll Road (TSTR) to connect raw material resources to the processing industry. Despite the benefits generated from highway project development, investment cost becomes the main issues that must be encountered besides land acquisition [11, 12]. The project requires 300 trillion rupiahs or equal to US$ 24 billion for 2700 km of highway development. The government of Indonesia has insufficient funds to accommodate all the required highway construction and thus should take other options by transferring part of the financing obligation to the private investors [13].

Researcher and academicians discussed that the project attractiveness was started by an optimum feasibility [14, 15]. TSTR has a low financial rate of return of about 7.79%. The figure relatively low compared to minimum attractiveness rate of return (MARR) from infrastructure/transportation investors' perspective more than 10% in the country. Consequently, companies hesitate to be involved when the project profoundly compromises their cash flow. Ref. [16] also supported the argument that highway project will succeed when the route is at a dense population and high-income corridor. Considering those problems, improving the feasibility by taking into account added value to the project becomes crucial.

The study will use value engineering (VE) approach in an attempt to improve project feasibility. VE has been implemented in various countries to improve the project or product development through a cost reduction or additional advantage by proposing significant benefits over cost [17, 18]. The United States has implemented VE approach during the past 30 years by including the method into the National Highway System (NHS) project that receives assistance with a total cost of $50,000,000 or more. VE is proven to improve transportation project performance and successfully saved 5.9% from the total cost of the project in 2015 [19].

The study contributes to the body of knowledge application of value engineering for highway development. The method offers optimum value that can attract investor in financing the project. The result of this study is also expected to be used for similar context not only in transportation but also for other value-added creations.

http://dx.doi.org/10.5772/intechopen.71778

2. Literature studies

Value engineering (VE) is a problem-solving technique that combines knowledge from different backgrounds organized in a team to conduct a thorough analysis and generate alternative solutions to targeted problems [20]. VE also defines as the multidisciplinary team attempt to evaluate the project by using particular standard through VE job plan for improvement. It generates alternatives and further investigates to meet the requirements of safety, quality, operations, maintenance, and aesthetics [21, 22].

Value engineering concept is broadly focused on how values can achieve a balance among time, cost, and quality. Value can be generated from four combinations. First, maintain the functions and quality while cost is reduced. Second, maintain cost while at the same time increase the functions and quality. Third, increase function and quality while reducing the cost. Last, increase functions and quality while reducing the cost.

Function as a primary element in value engineering aims to become foundation and catalyst of innovative solution. The function is categorized into the primary and secondary function. Basic function is a reason for why the system is created or basis for the existence of a product. On the other hand, the secondary function is an indirect use that aims to support the basic functions. Furthermore, the cost is the sum of expenses in developing, producing, and applying product or project and often called as life-cycle cost (LCC). Elements of LCC consist of investment cost, financing cost, operational cost, maintenance cost, cost of change, tax, and salvage value.

A fast diagram is a tool that shows logic diagram from functions of the item, subsystem or facility that illustrated through graphics and aims to test the reliability of identified functions [17]. There are four types of fast diagram models such as classical fast model, hierarchy function model, technical fast model, and customer-oriented fast model. The Classical fast model illustrated the interrelationship between functions in "HOW-WHY" logic. On the other hand, hierarchy function model used hierarchy graphic from functions listed vertically. Furthermore, the technical fast model is another form of classical fast model but adding "all the time" function, "one-time" function, and "same time" function or "caused by" function in the diagram. Last, customer-oriented fast model is developed to reflect customer as the one who determines value in function analysis process. This function adds supportive functions: attract users, satisfy users, assure dependability, and assure convenience. The study follows the technical fast model to achieve targeted outcomes of the project. The sample of the technical fast model can be seen in **Figure 1**.

In general, the diagram has six components such as scope line, critical path, critical path functions, higher level and basic function, secondary and causative function, and supporting function. The description of each component is as follows:

- Scope line

Functions placed between two vertical lines. In the technical FAST diagram, there are two scope line: left scope line and right scope line.

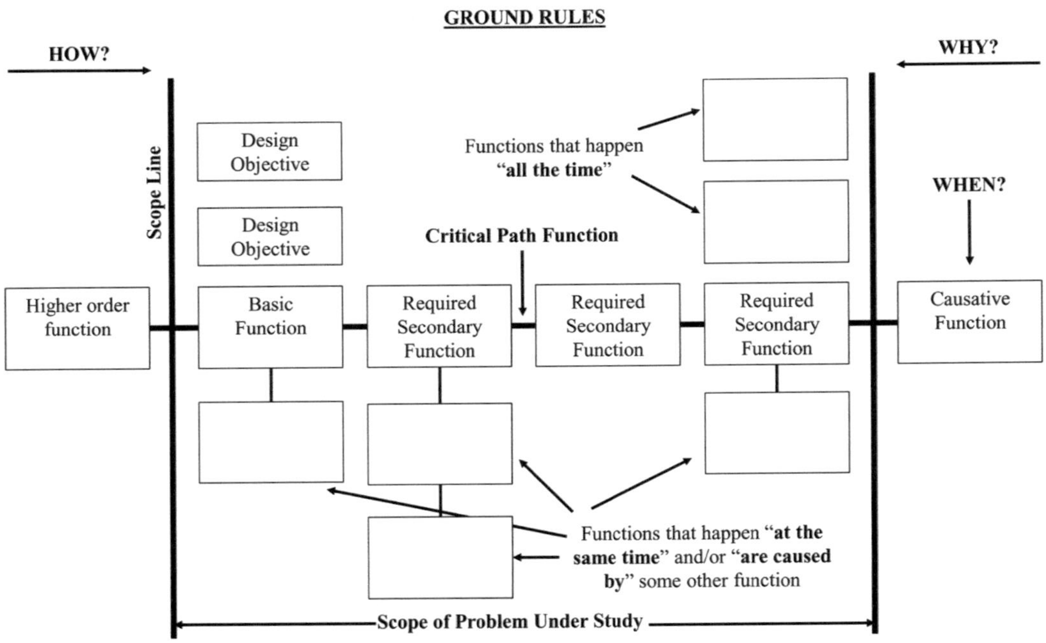

Figure 1. Technically oriented FAST diagram. Source: [23].

- Critical path

  Functions are placed on the critical path to fulfill users' expectation. While the critical path is determined, functions inside the scope of the project can be categorized into critical path functions and supporting functions. On the other hand, functions categorized outside the scope of the project are higher-order function and causative function.

- Critical path functions

  Functions at critical path have to place the order from the highest level to the lowest level or from left to right. The highest level function inside the scope of a project is called the basic function. Other functions at critical path inside the scope of a project are called required secondary functions.

- Higher level function and basic function

  Highest-order functions are placed outside the scope of the problem or left side of the left scope line. These functions require accomplishing user needs and becoming the reason why the project existed. Higher-order function is also responding "WHY" from a basic function, while in contrary, basic function answer "WHY" from the higher-order function.

- Secondary function and causative function

  The secondary function is all functions at the critical path with lower level from the basic function. The level of these functions is determined by HOW-WHY questions.

- Supporting functions

Supporting functions are functions inside scope line that are not located at critical path.

3. The Trans-Sumatera highway project

The Trans-Sumatera highway road project is one of mega-project infrastructures that is located in an island in the western part of the country called Sumatera Island. The project is estimated to stretch about 2700 km and forecasted require 330 trillion rupiahs. It shall connect eight mainland provinces out of 10 provinces in the area. The project is expected as part of Asian Highway network and includes on Master Plan on ASEAN Connectivity (MPAC).

The project is expected to generate benefits such as accelerate the logistic flow of goods and services distribution as well as open up opportunities about 97 thousand jobs during the project life cycle. It also contributes to the regional development indirectly and improves the property price in competitive ways [24].

Ministry of public works has been developing the route by taking flat topography on the western path of the island. The concept is divided into two corridors: a South-North corridor act as the primary corridor from Lampung province to Aceh province. On the other hand, West-East Corridor serves as the connecting corridor and comprises of three parts such as Palembang-Bengkulu, Pekanbaru-Padang, and Medan-Sibolga. The results show a different financial rate of return (FIRR) in each section and overall produce 7.79%. As the feasibility is below the minimum rate of return (MARR) of transportation companies, most investors tend to hold back their involvement. The government attempts to propose two bidding processes, and none of them were a success to financial close.

As the government tries to accelerate economic development, the construction of infrastructure is necessary for people's mobility and services. Thus, enabling added value to support financial and microeconomic growth should be created.

4. Developing conceptual design of Trans-Sumatera highway road

Conceptual design of the highway firstly was conducted by considering population, and economic landscape expressed by gross regional domestic product (GRDP) on the Sumatera island. Location quotient (LQ) was used to generate potential regencies and cities for the highway path. LQ analyzes and estimates the local economic basis and offers financial flexibility by taking into account industrial portfolio. Many sectors have been applying the method for a similar purpose from food, marine, and carbon emission [25–27]. It is expressed by a mathematical model as follows:

$$LQ = \frac{ei/e}{Ei/E} \tag{1}$$

where LQ is location quotient in a regency/district, ei is GRDP for i sector in a Regency/district, e is GRDP for the whole compared sector in a Regency/district, Ei is GRDP for i sector in the province, and E is GRDP for i whole compared sector in the province.

The LQ assess the potential sectors for each regency and cities on eight provinces on the island. The result will be considered as an input to determine the route of the project. The highway should connect regencies/districts that have a substantial impact on the whole island. Thus, a comprehensive route planning by taking into account potential commodity becomes significant to increase the project feasibility. The concept of the highway road can be seen in **Figure 2**. In total, the eight provinces in Sumatera island is connected. Overall, the entire length of the route is about 2427.15 km with 2102.10 km of lowlands, and 325.05 km of hills.

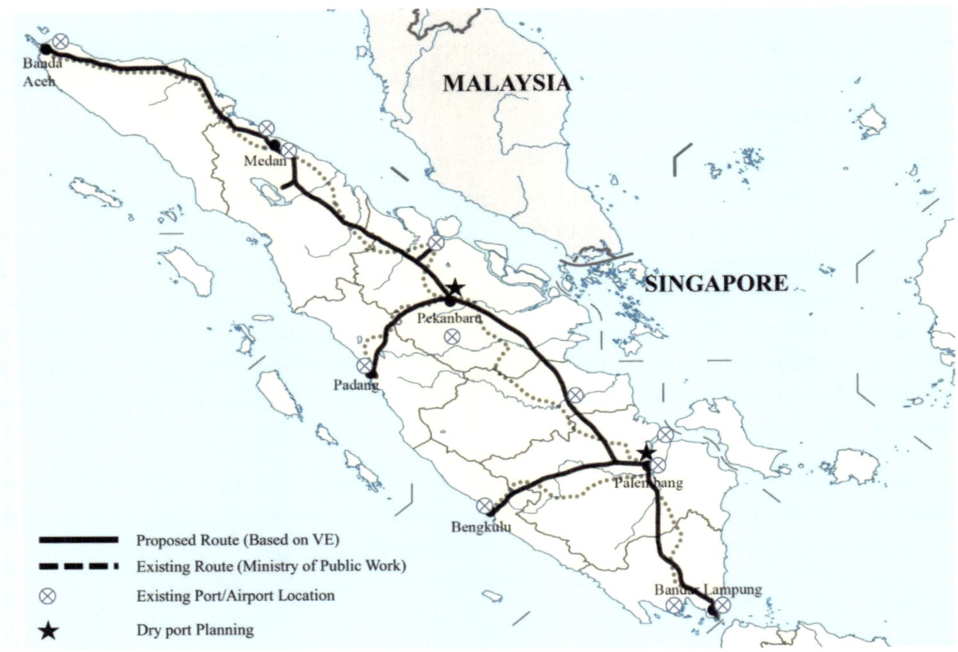

Figure 2. Conceptual design of Trans-Sumatera highway road. Source: [28].

5. Value engineering on Trans-Sumatera highway road

The value engineering in the study follows VE job plan and collaborates varied background in the process [29].

5.1. Information phase

In this stage, data gathering is practically being used to identify the project in more detail. The data and information are generated from journal articles, government and research reports, regulation, and policy and other related activities. The output was presented but not limited as follows:

- Project objectives

  The government has been an attempt to construct the project that stretches over Bakauheni to Aceh connecting not only cities but five airports and six main ports. The project is expected to support economic corridor on the island that mainly relies on natural resources, processing industries, and agriculture products. The project is also part of ASEAN and Asian Highway Network which was signed at 61st UN ESCAP on 26–28 April 2008, Shanghai.

- Project benefits

  The project aims to improve goods and services distribution especially among Sumatera and Java Island. As current artery road in the area is of over capacity and mostly of poor condition, the project construction will accelerate economic growth and reduce logistics cost. The project is also believed to allow employment during construction to the operational stage.

- Scope

  The project is in the conceptual design phase which assesses the route and potential added value to increase the project feasibility.

- Concept design existing

  The project comprises of two corridors that estimated to plan about 2700 km and forecasted require 330 trillion rupiahs. The central corridor is from Lampung province in the southern part of Sumatera to Aceh province in the northern part of Sumatera. On the other hand, connecting corridor consists of three sections: the city of Palembang at South Sumatera to Bengkulu, the city of Pekanbaru at Riau Province to the city of Padang at West Sumatera, and Medan-Sibolga both in North Sumatera.

5.2. Function analysis phase

The phase was conducted by identifying function by setting up the scope of the problem, order function, primary function, and others through a FAST diagram. It is a logical model tool that is used to identify, classify, develop, and select functions that contribute to higher value and benefit to the project development. The FAST diagram can be seen as shown in **Figure 3**.

5.3. Creativity phase

This phase evaluates the functional analysis in the previous stage by taking into account more detail potential on the project. It will update the previous FAST diagram model by including extended functions and extended processes. The revised concept design is shown in **Figure 4**.

5.4. Evaluation phase

The evaluation of the study is conducted by considering the project life cycle cost. The additional function that has been identified has to increase the feasibility, thus attracting private

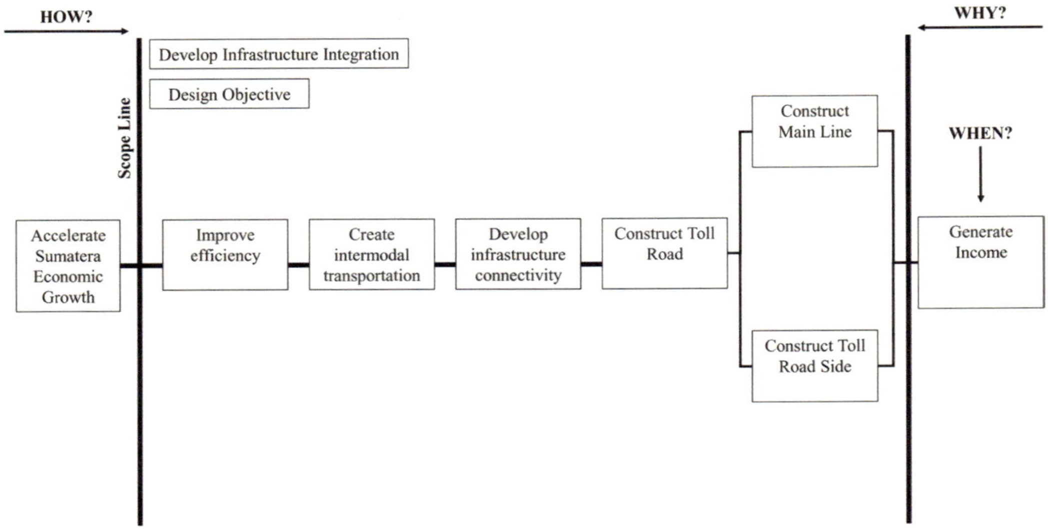

Figure 3. Initial FAST diagram.

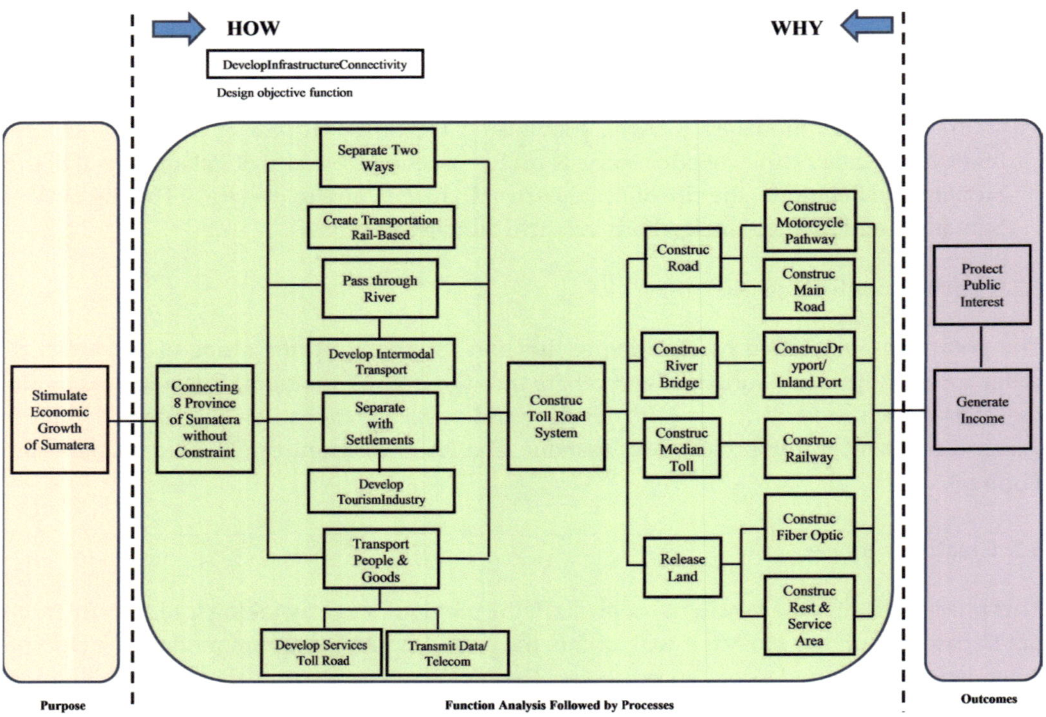

Figure 4. Revised FAST diagram. Source: [28].

investors. When the function was unable to generate financial benefits, the function can be justified through other advantages such as economic growth, job opportunity, and many others. The project life cycle cost is discussed in the following scenario.

6. Financial scenario

In evaluating the financial scenario, life-cycle cost approach is used by considering mid scenario for every identified function. It considers initial cost, operational and maintenance cost, and revenue. Several assumptions are proposed for estimation boundary which retrieved from Statistics Indonesia and other reliable sources. The discount rate is about 6.81%, inflation ranging from 1.63 to 5.95% depends on the sector, and project life cycle is about 40 years.

6.1. Single function

The new route of highway concept spans about 2527 km consist of 6 corridors and 22 sections. The initial cost is followed by the unit cost from the ministry of public works where the road on lowland is about 106.6 million rupiahs/km, and road on hilly terrain is about 164 million rupiahs/km. Revenue is generated by considering three scenarios of tariff and divided by main and connecting corridors. The scenario is as follows:

- Low scenario tariff: Main corridor = 150 rupiahs/km

  Connecting corridor = 300 rupiahs/km

- Mid scenario tariff: Main corridor = 300 rupiahs/km

  Connecting corridor = 300 rupiahs/km

- High scenario tariff: Main corridor = 300 rupiahs/km

  Connecting corridor = 600 rupiahs/km.

In term of low tariff scenario, 7 out of 22 scenarios generate negative net present value (NPV) that indicates the low internal rate of return (IRR). The average IRR at this scenario is about 7.25%. Middle tariff scenario produces 19 scenarios that show positive NPV. The average is higher than low tariff scenario as about 11.34%. Tebing Tinggi-Lima Puluh section is the lowest IRR about 4.88%. The low figure is caused by a high construction cost in contrast with low projected volume trip during 40 years of the operational stage. High tariff scenario proposes 20 scenarios with a positive NPV. The average IRR at this scenario is about 12.61%. Despite producing higher IRR, the scenario also shows sections with negative NPV such as Tebing Tinggi-Lima Puluh and Sigli-Banda Aceh. As financial engineering was unable to increase the sections into the targeted feasibility, additional functions were further proposed. The detail of the scenario can be seen in **Table 1**.

Highway section	Length	Tariff scenario 1		Tariff scenario 2		Tariff scenario 3	
		NPV	IRR (%)	NPV	IRR (%)	NPV	IRR (%)
Pekanbaru-Medan (main corridor)	**547.2**	**Tariff IDR 150/km**		**Tariff IDR 300/km**		**Tariff IDR 300/km**	
Bakauheni-Bandar Lampung	85.8	Negative	4.31	Positive	8.55	Positive	8.55
Bandar Lampung-Metro-Kota Bumi	102.3	Positive	11.92	Positive	17.32	Positive	17.32
Kota Bumi-Kayu Agung	185.3	Positive	7.21	Positive	12.32	Positive	12.32
Kayu Agung-Palembang	63.8	Positive	7.67	Positive	13.04	Positive	13.04
Palembang-Pangkalan Balai-Sekayu	110	Positive	11.30	Positive	18.63	Positive	18.63
Palembang-Pekanbaru (main corridor)	**494.8**	**Tariff IDR 150/km**		**Tariff IDR 300/km**		**Tariff IDR 300/km**	
Sekayu-Meranti Panjang-Jambi	151.5	Negative	2.87	Positive	7.37	Positive	7.37
Jambi-Rengat	184.9	Positive	7.76	Positive	13.17	Positive	13.17
Rengat-Pangkalan Kerinci-Pekanbaru	158.4	Positive	7.03	Positive	11.99	Positive	11.99
Pekanbaru-Medan (main corridor)	**581.4**	**Tariff IDR 150/km**		**Tariff IDR 300/km**		**Tariff IDR 300/km**	
Pekanbaru-Pematang Duri-Dumai	133.2	Positive	10.22	Positive	16.55	Positive	16.55
Pematang Duri-Rantau Prapat	191	Positive	6.86	Positive	11.99	Positive	11.99
Rantau Prapat-Pematang Siantar-Raya	135	Negative	5.39	Positive	10.35	Positive	10.35
Pematang Siantar-Tebing Tinggi-Medan	93.08	Negative	10.90	Positive	17.60	Positive	17.60
Tebing Tinggi-Lima Puluh	29.12	Negative	1.57	Negative	4.88	Negative	4.88
Banda-Aceh Medan (main corridor)	**480.2**	**Tariff IDR 150/km**		**Tariff IDR 300/km**		**Tariff IDR 300/km**	
Medan-Langsa	118	Positive	9.68	Positive	16.16	Positive	16.16
Langsa-Lhokseumawe	136.8	Negative	6.72	Positive	12.27	Positive	12.27
Lhokseumawe-Sigli	145.5	Negative	3.61	Positive	8.78	Positive	8.78
Sigli-Banda Aceh	79.9	Negative	2.61	Negative	6.58	Negative	6.58

Highway section	Length	Tariff scenario 1		Tariff scenario 2		Tariff scenario 3	
		NPV	IRR (%)	NPV	IRR (%)	NPV	IRR (%)
Pekanbaru-Padang (connecting corridor)	**212.9**	**Tariff IDR 300/km**		**Tariff IDR 300/km**		**Tariff IDR 600/km**	
Pekanbaru-Bangkinang	50.4	Positive	7.06	Positive	7.06	Positive	11.96
Bangkinang-Paya Kambuh	72.5	Positive	7.78	Positive	7.78	Positive	12.97
Paya Kambuh-Batusangkar-Padang	90	Positive	12.41	Positive	12.41	Positive	19.65
Palembang-Bengkulu (connecting corridor)	**211.1**	**Tariff IDR 300/km**		**Tariff IDR 300/km**		**Tariff IDR 600/km**	
Sekayu-Lubuklinggau	116.1	Positive	9.55	Positive	9.55	Positive	15.37
Lubuklinggau-Bengkulu	95	Negative	5.13	Negative	5.13	Positive	9.81
Average IRR		**Average IRR**	**7.25**	**Average IRR**	**11.34**	**Average IRR**	**12.61**

Table 1. Life cycle cost analysis of highway function.

6.2. Multifunction

The concept integrates highway with other potential function such as dry port, fiber optic, tourism, rest area, and services, as well as railway and motorbike integration. Each function is investigated with the initial cost, operation, and maintenance cost and revenue for 40 years of operation. Rest area and tourism development were identified as the two most contributing functions in term of revenue. It was followed by railway integration, dry port, motorbike integration, and fiber optic. The projected revenue for each additional function can be seen in **Figure 5**.

Compared to the single function, the highway with multifunction concept generates more revenue for the project. Low tariff in multifunction concept produces 200 billion rupiahs in 2060 compared to 160 billion rupiahs for the single function. Moreover, the mid tariff on multifunction concept surpasses the revenue from mid tariff and higher tariff scenario of the single-function concept. It produces 350 billion rupiahs in 2060 compared to 330 billion rupiahs and 290 billion rupiahs, respectively. The projected revenue for multifunction and single function can be seen in **Figure 6**.

Life cycle cost in multifunction simulated through 8 alternative scenarios. Each alternative differed from the active function and elaborated with IRR, NPV, and payback period (PbP). Alternative A recommends including all additional functions, while some others inactivated the additional function. The simulation can be seen in **Table 2**.

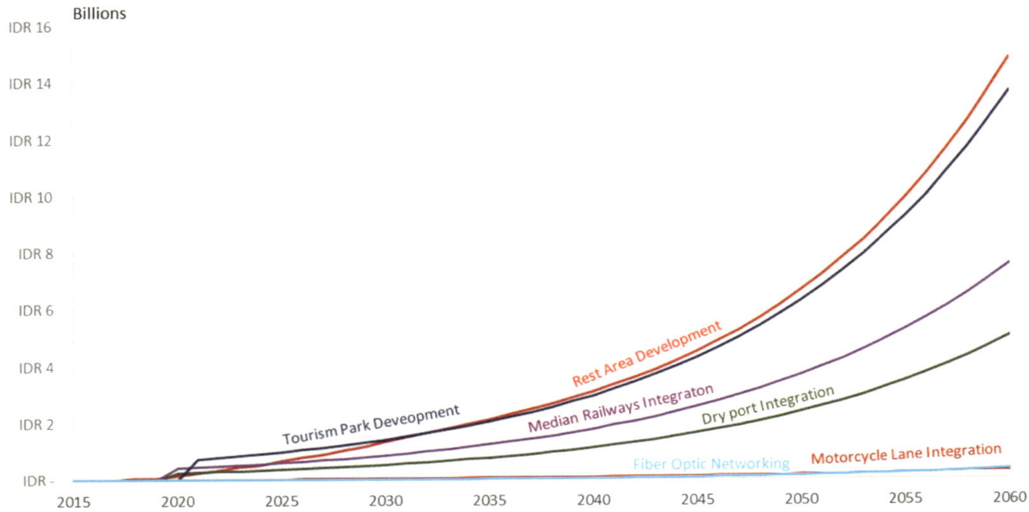

Figure 5. Projected revenue for each additional function on the conceptual design. Source: [28].

In terms of the low tariff, alternative E generates higher IRR with the lowest PbP about 8.30% and 34.39%. Despite lower IRR and higher PbP, alternative A provides more NPV compared to the alternative E. Similar issue also shows for mid and high tariff where alternative E proposes higher IRR and PbP, yet alternative A suggests higher NPV.

Alternative E suggests excluding tourism for the project. However, the tourism acts as one of the highest contributor to project revenue. With such valuable attractiveness, excluding the sector is no longer an option. As a result, alternative A that includes all additional functions is proposed as the best scenario for the project development. The detailed comparison for each alternative in terms of the financial feasibility is shown in **Table 3**.

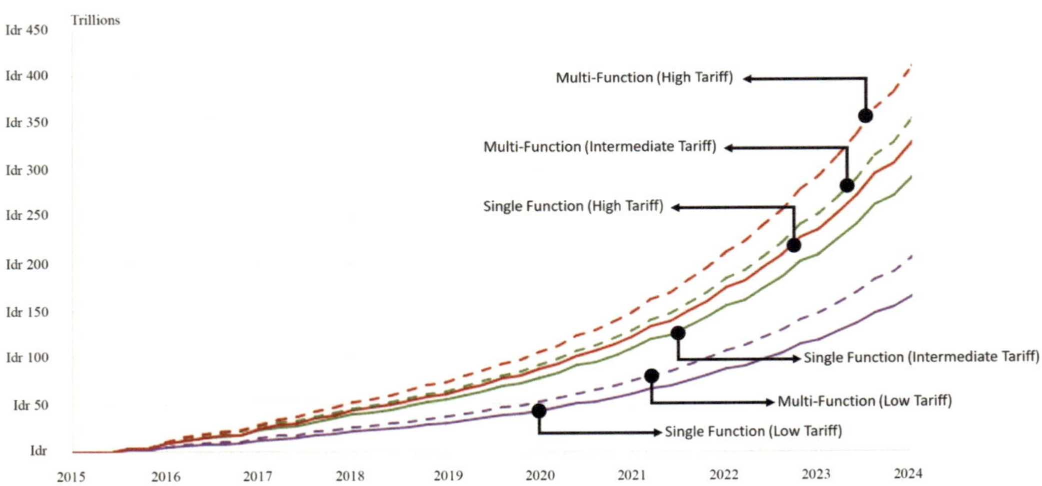

Figure 6. Projected revenue comparison for multifunction and single function. Source: [28].

Alternative	Highway road	Rest area	Tourism	Dry port	Railways	Motorcycle	Fiber optic
A	√	√	√	√	√	√	√
B	√	√	√	√	√	x	x
C	√	√	√	x	x	√	√
D	√	√	√	x	x	x	x
E	√	√	x	√	√	√	√
F	√	x	√	√	√	√	√
G	√	x	x	√	√	x	x
H	√	x	x	x	x	√	√

Table 2. Life cycle cost simulation for multifunction conceptual design.

Alt	Low tariff			Intermediate tariff			High tariff		
	NPV (million IDR)	IRR (%)	PBP	NPV (million IDR)	IRR (%)	PBP	NPV (million IDR)	IRR (%)	PBP
A	83,743,843	8.29	34.47	398,199,357	12.71	20.72	501,149,460	13.77	19.62
B	83,609,123	8.29	34.46	397,025,087	12.70	20.45	498,935,639	13.76	18.65
C	76,617,476	8.22	34.76	379,764,988	12.66	20.52	471,407,089	13.63	18.83
D	76,482,757	8.22	34.76	378,590,718	12.65	20.52	469,193,268	13.61	18.88
E	78,811,227	8.30	34.39	380,774,195	12.84	20.22	471,231,753	13.82	18.59
F	72,121,410	8.13	35.27	375,411,362	12.58	20.74	467,195,903	13.57	20.00
G	67,054,074	8.13	35.35	356,811,930	12.71	20.54	435,064,376	13.60	19.02
H	60,062,427	8.04	35.36	339,551,832	12.66	20.59	407,535,826	13.45	19.19

Table 3. Scenario on eight alternative for multifunction conceptual design.

7. Public-private partnership

The scheme will evaluate division of responsibility between the government and private on the life cycle of the project from initial cost, operational and maintenance cost, as well as revenue. The financing scheme consists of three scenarios:

- Scenario A: Government share 60% and private 40% for the initial cost.
- Scenario B: Government share 50% and private 50% for the initial cost.
- Scenario C: Government share 40% and private 60% for the initial cost.

Each scenario is followed by cost sharing on operational and maintenance cost where the private contributes 60, 50, and 40%. Revenue will also be shared where the private obtains from 80 to 50%.

In evaluating the best scenario, three considerations are being used from the comparison of IRR and MARR, revenue weighting score, and weighting score of operational and maintenance cost. Revenue weighting score for private shall be the lowest. When IRR reach expected MARR with the lowest revenue, calculation using higher revenue is not necessary. A similar scenario is also conducted on the O&M cost. When the highest O&M cost contributes to reaching IRR over MARR, then lower cost is no longer needed.

7.1. Scenario A

The scenario A consists of 12 initial scenarios from A.1 to A.12. Firstly, parameter about IRR minimum is conducted, and 6 scenarios are passed. Second, revenue weighting score for private to gain expected MARR (12%) is evaluated, and it generates three available scenarios from A.2, A.6, and A.10. Last, weighting score of operational and maintenance cost produces two most potential scenarios consisting of A.6 and A.10.

Both scenarios regulate private responsibility on initial cost and operational and maintenance cost as well as private's revenue from the project. A.6 scenario produces 12.48% of the internal rate of return (IRR) with a positive NPV. The private on initial cost and operational and maintenance cost shall contribute about 60 and 50%, respectively. In this scenario, private will obtain 70% of revenue from the project. On the other hand, A.10 scenario produces IRR for about 12.31% with a positive NPV. In the scenario, private has different contribution percentage on operational and maintenance cost, that is, about 60%.

7.2. Scenario B

The scenario B consists of 12 initial scenarios from B.1 to B.12. Similar treatment was also conducted as scenario A. The scenario generates two most potential scenarios B.7 and B.11.

B.7 scenario produces 12.63% with positive NPV. The scenario suggests that the private contributes 50% of initial cost and operational and maintenance cost. They reserve the right to gain 60% of revenue. Moreover, B.11 scenario produces close IRR to B.7 about 12.43% and a positive NPV. Private obtains 60% of revenue by contributing 50% of the initial cost and 60% of operational and maintenance cost.

7.3. Scenario C

The scenario C consists of 12 initial scenarios from C.1 to C.12. Similar treatment was also conducted as scenario A and scenario B. The scenario generates two most potential scenarios C.8 and C.12.

C.8 scenario produces 12.86% with positive NPV. The scenario regulates private responsibility and rights for initial cost, operational and maintenance cost as well as revenue by 40, 50 and 50%, correspondingly. C.12 generates 12.60% of IRR and a positive NPV. The private should invest 40% of initial cost and 60% of operational and maintenance cost. They will gain 50% of revenue from the investment.

http://dx.doi.org/10.5772/intechopen.71778

7.4. The proposed scenario

As 6 scenarios have been generated, then the most optimal financing scheme by the division of responsibility between government and private shall be conducted. All the scenarios show similar IRR about 12%. The comparison for each scenario can be seen in **Table 4**. It shows the investor responsibility for initial cost to revenue.

From the 6 available scenarios, the C.8 scenario is recommended for the division of responsibility between the public and private partnership. In public perspective, government contributes sunk cost about 60% from the project but gains revenue about 50% which can be used as a capital investment to support other infrastructure projects. For private, they contribute higher operational and maintenance cost compared to other scenarios and share equal income for the revenue. As the initial cost for private is only at 40%, the IRR produces competitive value. The proposed scheme aims to benefit public interest without compromising private income from the project.

Scenario	Initial cost (%)	OM cost (%)	Revenue (%)	IRR (%)
A.6	60	50	70	12.48
A.10	60	60	70	12.31
B.7	50	50	60	12.63
B.11	50	60	60	12.42
C.8	40	50	50	12.86
C.12	40	60	50	12.60

Table 4. Comparison for optimal scenario.

8. Institutional scheme

The financing scheme produces an optimum scenario which increases the IRR value from 10.59 to 12.86%. As the project uses more than one function, it is argued that a consortium which accommodates multidiscipline sector should be established. The project is consisted of nine main stakeholders: the joint venture company, investor, consultant, insurance institution, national bank, multilateral bank, contractor, operator, and user.

The joint venture company consists of six different sectors such as transportation, telecommunication, tourism, port, railway, and industry. They should collaborate with the government to institute a business entity which shall manage the project. Investor is supported by a national bank and shares the cost with companies in the joint venture. On the other hand, the multilateral bank shall support government through soft loan. The detail institutional scheme is shown in **Figure 7**.

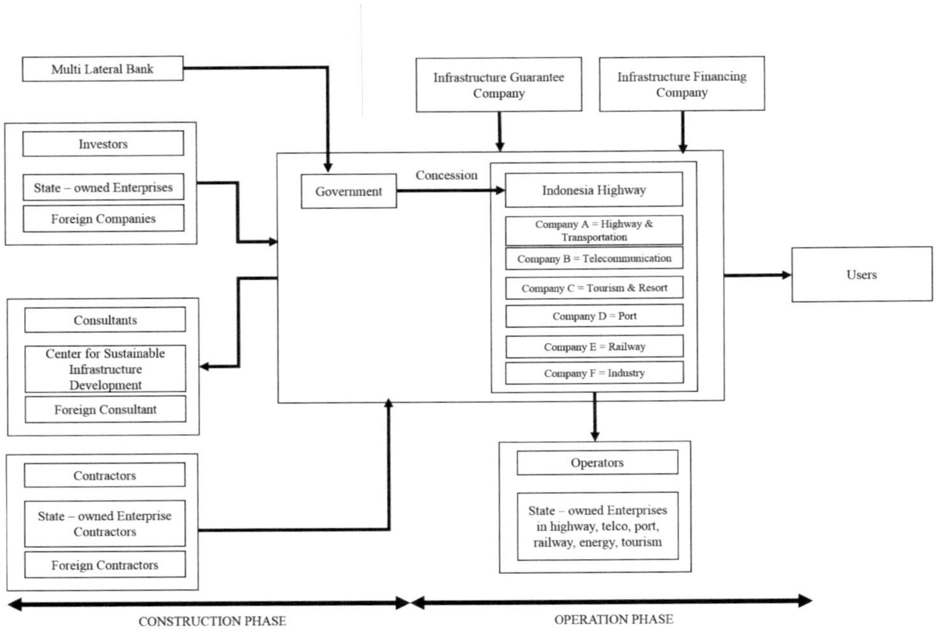

Figure 7. Institutional scheme on the project.

9. Conclusion

Value engineering (VE) has successfully improved the highway project by creating innovation and additional function. Value for money project reaches through the transformation of a single-function project into a multifunction project. The VE process generates dry port, fiber optic, tourism, rest area, and services, as well as railway and motorbike integration to meet the innovative conceptual design of highway development.

Compared to the single function project that produces 7.79%, the multifunction of TSTR project proposes a rate of return of about 10.59%. The study elaborates financing scheme that is suitable for government and business through public-private partnership. The system proposed as build operate sharing transfer (BOST) where government play roles on each project life cycle. Government contributes 59% in the initial stage and 40% in the operational and maintenance stage, and obtains 21% from the projected revenue. The public-private partnership scheme increases rate of return from 10.59 to 12.76% and a positive net present value (NPV). The study also generates institutional scheme that regulates related stakeholders from investors, joint venture structure, consultant, to the contractors for a division of responsibility for the project life cycle.

Acknowledgements

This research is supported by research grants from the Ministry of Research, Technology and Higher Education and Universitas Indonesia.

Author details

Mohammed Ali Berawi

Address all correspondence to: maberawi@eng.ui.ac.id

Faculty of Engineering, Universitas Indonesia, Indonesia

References

[1] Bel G. Infrastructure and nation building: The regulation and financing of network transportation infrastructures in Spain (1720-2010). Business History. 2011;**53**(5):688-705

[2] Berawi MA, Zagloel TY, Miraj P, Berawi AB, Titaheluw W, Karim SBBA. Developing road infrastructure route planning: Increasing feasibility of toll road project. Makara Journal of Technology. 2016;**20**(3):147-152

[3] Meijers E, Hoekstra J, Leijten M, Louw E, Spaans M. Connecting the periphery: Distributive effects of new infrastructure. Journal of Transport Geography. 2012;**22**:187-198

[4] Otsuka A, Goto M, Sueyoshi T. Industrial agglomeration effects in Japan: Productive efficiency, market access, and public fiscal transfer. Papers in Regional Science. 2010; **89**(4):819-840

[5] Li Z, Hensher DA. Toll roads in Australia: An overview of characteristics and accuracy of demand forecasts. Transport Reviews. 2010;**30**(5):541-569

[6] World Bank. Population of Country - Indonesia [Internet]. 2016. Available from: http://data.worldbank.org/indicator/SP.POP.TOTL?locations=ID [Accessed: July 2017]

[7] Toll Road Regulatory Agency. Operated Toll Road in Indonesia. Ministry of Public Works; 2017

[8] Ministry of Land, Infrastructure, Transport and Tourism. Expressway in Japan. 2015 www.mlit.go.jp/road/road_e/pdf/ROAD2014web.pdf [Retrieved from: 9 November 2017]

[9] Cho SM. Road safety facilities in Korea. Expert Group Meeting on Asian Highway; 3, 4 Oct 2016; Bangkok, Thailand. 2016. http://www.unescap.org/sites/default/files/3.4%20Republic%20of%20Korea_Mr.%20Sungmin%20Cho.pdf [Retrieved from: 9 November 2017]

[10] Malaysia Highway Authority. Malaysian Roads. 2012. https://www.piarc.org/ressources/documents/1216,road-network-in-malaysia-v2.pdf retrieved [9 November 2017]

[11] Kim E, Bae Y. Economic contribution of the private sector on financing and operation of a highway: Financial computable general equilibrium. Journal of Transport Economics and Policy (JTEP). 2015;**49**(3):397-414

[12] Kumar L, Jindal A, Velaga NR. Financial risk assessment and modelling of PPP based Indian highway infrastructure projects. Transport Policy. 2017. DOI: 10.1016/j.tranpol.2017.03.010

[13] Khmel V, Zhao S. Arrangement of financing for highway infrastructure projects under the conditions of public–private partnership. IATSS Research. 2016;**39**(2):138-145

[14] Berawi MA, Berawi ARB, Prajitno I, Nahry N, Miraj P, Abdurachman Y, Tobing E, Ivan A. Developing conceptual design of high-speed railways using value engineering method: Creating optimum project benefits. International Journal of Technology. 2015; **6**(4):670-679

[15] Liu J, Love PE, Smith J, Regan M, Davis PR. Life cycle critical success factors for public-private partnership infrastructure projects. Journal of Management in Engineering. 2014; **31**(5):04014073

[16] Muller RH. Examining Toll Road Feasibility Studies. Public Works Financing, NJ USA; 1996

[17] Woodhead RM, Berawi MA. An alternative theory of idea generation. International Journal of Management Practice. 2008;**3**(1):1-19

[18] Berawi MA, Susantono B. Developing conceptual design of mega infrastructure project: creating innovation and added value. Value World. 2012;**35**(1):12-20

[19] Federal Highway Administration, Department of Transportation. Value Engineering Final Rule. 2017. https://www.fhwa.dot.gov/ve/

[20] Dell'Isola A. Value Engineering: Practical Applications… for Design, Construction, Maintenance and Operations. Vol. 35. RSMeans; 1997

[21] Younker D. Value Engineering: Analysis and Methodology. CRC Press, Boca Raton, Florida, USA; 2003

[22] Shen Q, Liu G. The selection of benchmarking partners for value management: an analytic approach. International Journal of Construction Management. 2007;**7**(2):11-22

[23] Snodgrass TJ, Kasi M. Function Analysis: The Stepping Stones to Good Value. University of Wisconsin-Madison, Wisconsin, USA; 1986

[24] Berawi MA, Zagloel TY, Miraj P, Mulyanto H. Producing alternative concept for the Trans-Sumatera toll road project development using location quotient method. Procedia Engineering. 2017;**171**:265-273

[25] Morrissey K. Producing regional production multipliers for Irish marine sector policy: A location quotient approach. Ocean & Coastal Management. 2014;**91**:58-64

[26] Trappey AJ, Trappey CV, Liu PH, Hsiao CT, Ou JJ, Chen KW. Location quotient EIO-LCA method for carbon emission analysis. In: Concurrent Engineering Approaches for Sustainable Product Development in a Multi-Disciplinary Environment. London: Springer; 2013. pp. 367-377

[27] Leslie TF, Frankenfeld CL, Makara MA. The spatial food environment of the DC metropolitan area: Clustering, co-location, and categorical differentiation. Applied Geography. 2012;**35**(1):300-307

[28] Berawi MA, Zagloel TM, Berawi ARB, Abdurachman Y. Feasibility analysis of Trans-Sumatera toll road using value engineering method. International Journal of Technology; 2015;**6**(3):388-399. DOI: 10.14716/ijtech.v6i3.1475

[29] Miles LD. Techniques of Value Analysis and Engineering. Miles Value Foundation, USA; 2015

Chapter 6

Nondestructive Tests in Highway Engineering

Gunawan Handayani

Additional information is available at the end of the chapter

http://dx.doi.org/10.5772/intechopen.71615

Abstract

In highway engineering, one needs to know the information about the thickness and elastic modulus of each pavement layer. The thickness and elastic modulus of each pavement component then become input to elasticity analysis or finite element computation to calculate the service life of the highway. An application was Multichannel Analysis of Surface Waves (MASW) measurement on highway in North Jakarta. The measurement was carried out on highway traffic, because we were not allowed to stop traffic of one of the busiest roads in Jakarta. The street is heading to Tanjung Priok, the port of Jakarta. However, we acquired fairly good data by applying stacking method. The number of stacking was 10 (ten) to overcome traffic noise. After inversion, we came up with the result of MASW measurement of the pavement. The result of MASW measurement in terms of elastic modulus and thicknesses of pavement layer becomes important input of finite element analysis to compute fatigue damage of pavement components.

Keywords: Multichannel Analysis of Surface Waves, MASW, surface waves, pavement

1. Introduction

In highway engineering, it is required to know the information about the thicknesses of pavement components and values of elastic modulus of pavement components. It is preferred to have nondestructive methods than destructive method (drilling method). The nondestructive methods are quicker to carry out than the destructive methods. More importantly, the pavement is not disturbed, i.e., no holes are required to be patched. Surface waves are employed in this nondestructive method. The method utilizes surface waves to determine the thickness and shear wave velocity of each layer. This chapter describes the procedure and methodology of surface wave method using multichannel recording. This method is popularly known as Multichannel Analysis of Surface Waves (MASW). This chapter further discusses the application of MASW method to busy highway in Jakarta. The high traffic noise was overcome by a recording technique known as stacking technique. This technique added to the fact that surface wave has the largest amplitude than other body waves (P and S waves) [1] resulted in good field data.

2. Methodology

The nondestructive method employed here is carried out by means of generating seismic waves on the surface. If one generates an impulsive source at the surface, it propagates seismic body waves (direct, refracted, reflected, scattered, guided, and air waves) and seismic surface waves (fundamental, higher modes, and scattered). The frequency content of the impulsive signal is very wide from nearly zero Hz until infinite Hz as can be seen from Fast Fourier Transforms computation below (**Figure 1**):

The pavement structure has strong layer in the upper layer and softer layer underneath. In this situation, the Snell's law governs that the wave would be refracted down if it meets interface between strong layer and soft layer. This is the main reason, the body wave method, i.e., refraction method cannot be applied (**Figure 2**).

By hammer blow, one generates impulsive signal and propagates body waves and surface waves into the pavement system. If one sets up an array of receiver sensors, i.e., geophones or accelerometers, he acquires shot gather, which is a recorded time series file of receiver sensors. An example of shot gather is shown in **Figure 3**.

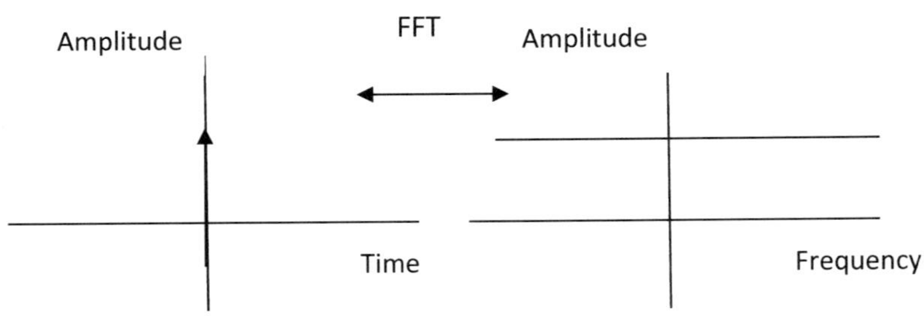

Figure 1. FFT relationship between time duration of a signal and its spectrum.

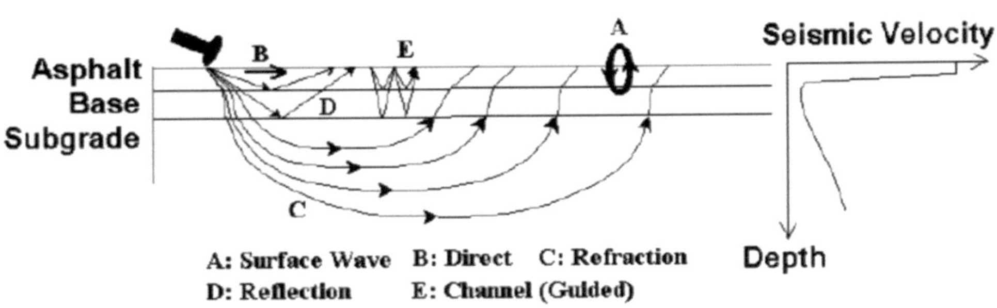

Figure 2. Generalized structure of an Asphalt system [2].

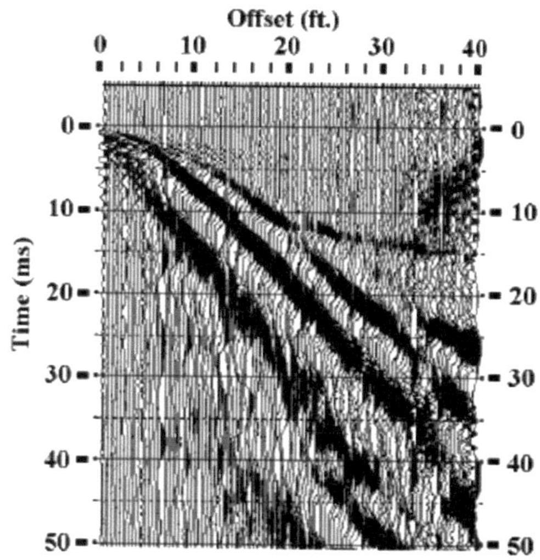

Figure 3. An example of shot gather obtained from hammer blow on the surface.

As it can be seen from the shot gather, the dominant event of arrival waves is surface waves, because their amplitude is very large compared to body waves (P or S waves). For the surface waves, people usually assume they contain fundamental mode. They actually contain higher modes too. However, up to now, the inversion algorithms usually ignore the higher modes. As hammer hit the surface, the surface waves propagate through the pavement system. The high frequency component (the small wavelength component) propagates through the upper thinnest pavement layer, whereas the lower frequency component (the larger wavelength) will propagate and vibrate through the deeper pavement layer. As the result of this process, there is velocity discrepancy between the high-frequency component, which travels with velocity of first layer, and lower frequency component, which travels with velocity of deeper layer (material). This velocity discrepancy is called dispersion, which means velocity as function of frequency (wavelength) (**Figure 4**) [3].

The shot gather from the hammer blow can be considered as time series:

$$f(t) = a_0 + \sum_{n=1}^{\infty}\left(a_n \cos\frac{2\pi nt}{T} + b_n \sin\frac{2\pi nt}{T}\right) \tag{1}$$

2.1. Forward modeling

If we have soil model on the left, the time series (shot gather) would be on the right (**Figures 5–7**):

The field records we acquired from the field then undergo data processing with the flow:

1. Compute Fast Fourier Transform (FFT) to the shot gather f(x, t):

$$F(\omega) = \frac{1}{2\pi}\int_{-\infty}^{\infty} f(t) \,.\, e^{-i\omega t}\, dt \tag{2}$$

(a)

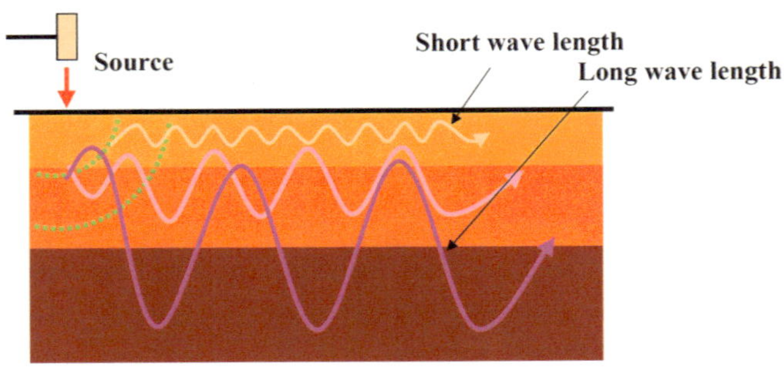

(b)

Figure 4. (a) Hammer blow as the source of energy. (b) Hammer blow would propagate many wavelength (frequencies).

and we obtain F(x, ω) record, and:

2. Redo the Fast Fourier Transform (FFT) to F(x, ω):

$$F(c,\omega) = \int_{-\infty}^{\infty} F(x,\omega) \, . \, e^{i\omega\frac{x}{c}} dx \tag{3}$$

where c = the phase velocity of the surface wave; ω = the angular frequency; x = distance of the sensor.

The whole process can be described in the following flow chart (**Figures 8–11**) [5]:

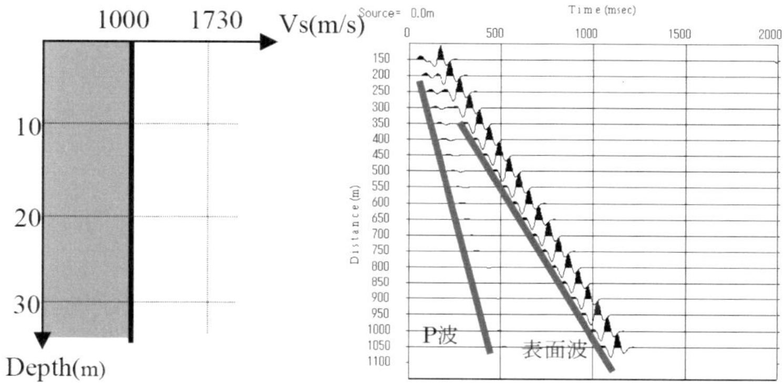

Figure 5. A soil model with velocity of 1000 m/s [4].

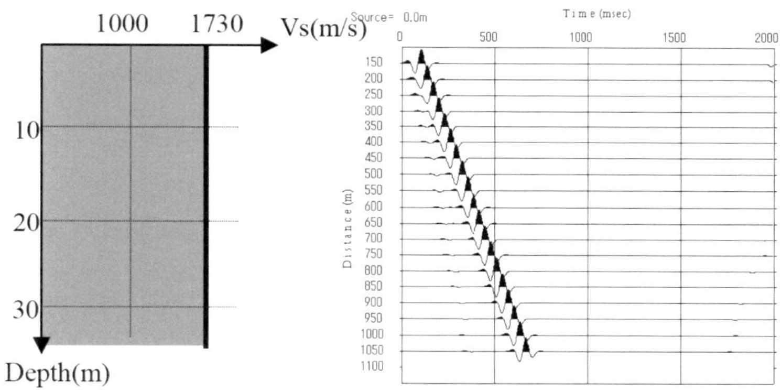

Figure 6. A soil model with velocity of 1730 m/s [4].

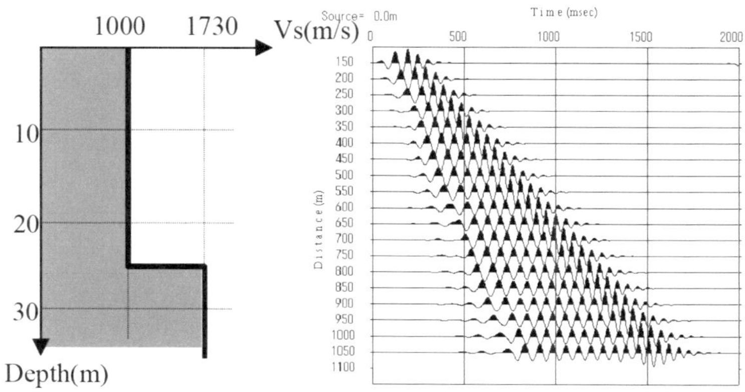

Figure 7. Combination of soil model with velocity of 1000 m/s and of 1730 m/s [4].

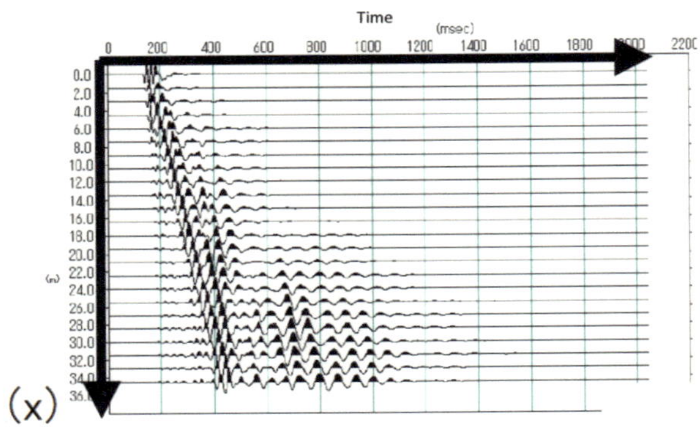

Figure 8. The shot gather.

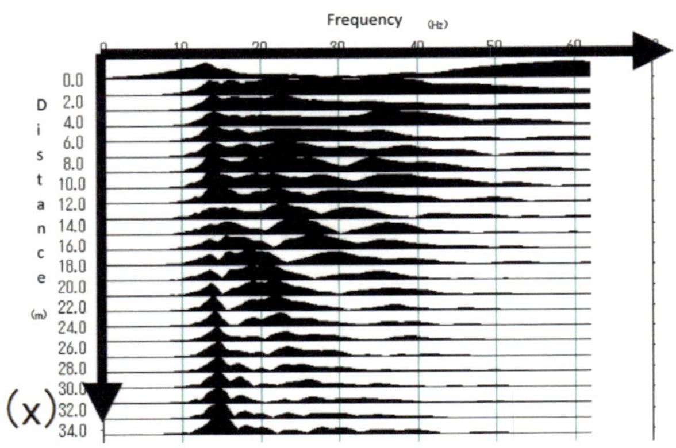

Figure 9. After FFT computation one obtains F(x, w).

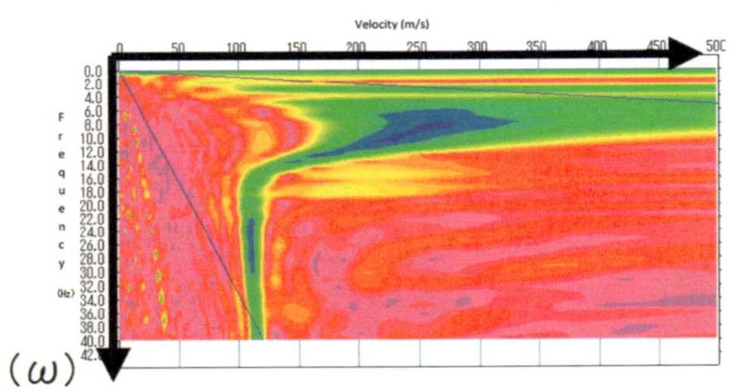

Figure 10. Finally one obtains spectrum velocity with respect to angular frequency F(c, ω) [4].

At this point, we analyze the waves, which are received by geophones from the spectrum F(c, ω), i.e.:

C: surface waves (fundamental mode).

B: surface wave (Higher mode).

A: body waves.

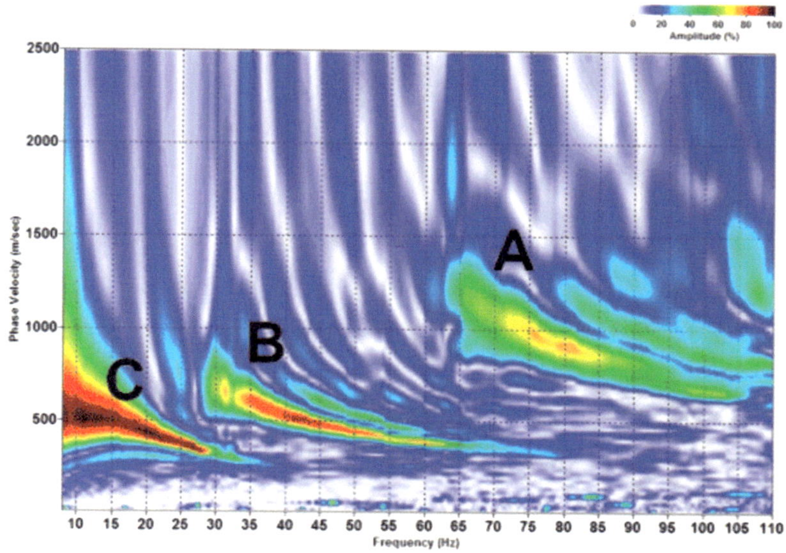

Figure 11. Pattern recognition of wave type in spectrum velocity with respect frequency [6].

For inversion purpose, we pick the surface wave (fundamental mode) only which is characterized by larger amplitude and lower velocity.

To record the surface waves, we employ geophones sitting on aluminum platform called land streamer, so there is no need to drill the surface of the pavement [7].

After picking the fundamental mode of the surface waves, one has the field dispersion curve to be inverted by utilizing commercial inversion program to come up with shear velocity profiles, i.e., velocity of shear wave with respect to depth. The inversion programs themselves are of many different kinds. We only focus on least square inversion program (**Figure 12**) [8].

2.2. The highway application

As an illustration of the process above, let us look at the real application, i.e., MASW (Multichannel Analysis of Surface Waves) measurement on the busy street in North Jakarta. The traffic was not allowed to be stopped, since it could disrupt the economic arteries of Jakarta and the street was heading to the Tanjung Priok, Jakarta's port. Therefore, we carried out the measurement side by side with busy traffic. The equipments employed are seismic data logger seistronix 24 bit, OYO geophones 4.5 Hz, and land streamer. The field parameters were (a) near offset (distance between the weight drop blows and the first geophone) was 18 m, (b) inter distance among geophones was 3 m, and (c) number of geophones was 12 geophones (**Figure 13**).

Even though a lot of street noise, we managed to acquire a fairly good data by applying stacking technique, which is basically we add up the records from the same source offset distance:

$$a_t = \frac{1}{N}\sum_{i=1}^{N} S_i \tag{4}$$

where S_i = amplitude of ith record; N = number of repetition; a_t = amplitude of stacked record.

Figure 12. A land streamer.

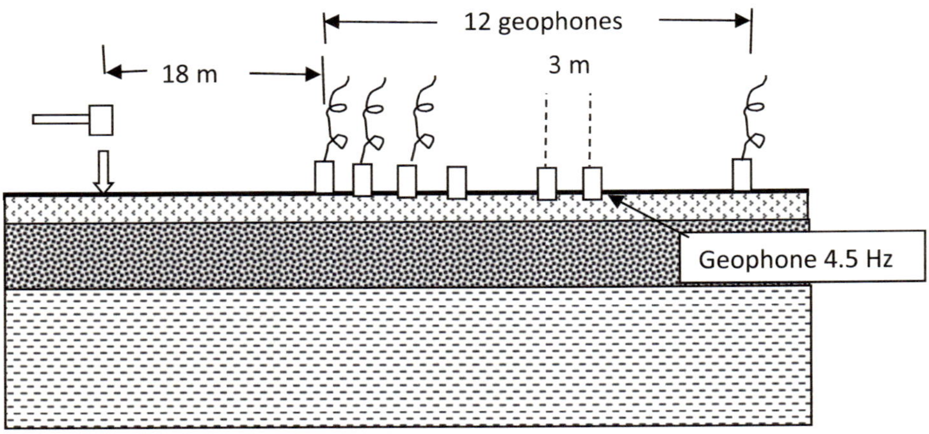

Figure 13. Field configuration of MASW measurement along with field measurement parameters [9].

Using this technique, random noise will add up to the same amplitude and after dividing it by number of repetition becomes one amplitude over number of repetition, whereas adding up coherence signal would result in number of repetition times original amplitude. After dividing it by number of repetition, we obtain initial amplitude. In this method, we suppress the traffic noise as one over number of repetition (**Figure 14**).

After stacking method, this shot gather was acquired as follows (**Figures 15** and **16**):

The FFT computation was carried out and resulted in spectrum F(c, ω) as follows (**Figure 17**):

Figure 14. MASW data acquisition using land streamer on busy street.

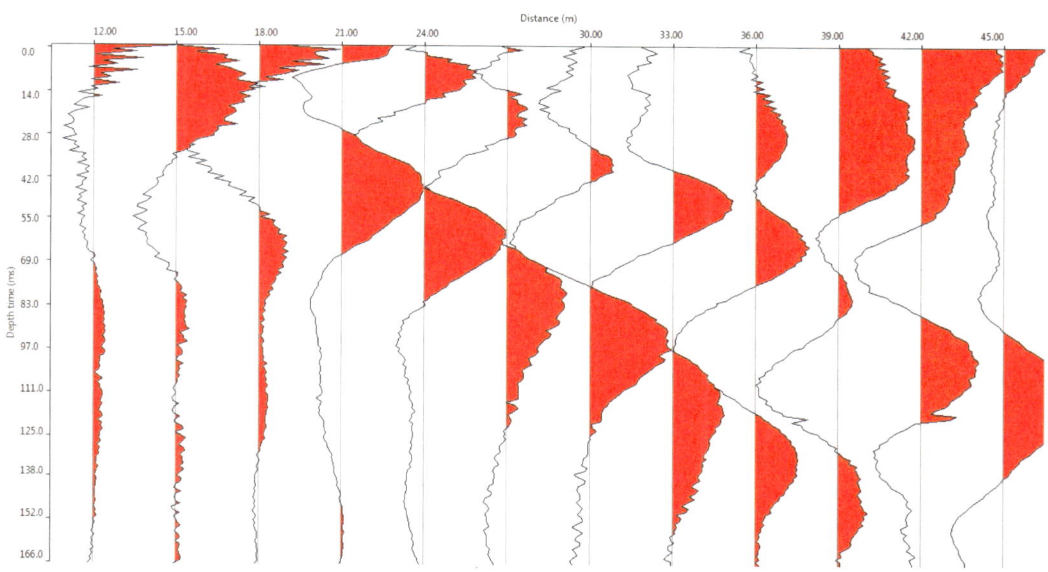

Figure 15. The record of field shot gather.

The fundamental mode corresponding to surface layer (the one which is extending to high frequencies) was picked and come up with the dispersion curve to be inverted/matched using commercial software "*SWAN*" as follows (**Figures 18** and **19**):

Figure 16. The weight drop used to generate vibration in repeat fashion (stacking method).

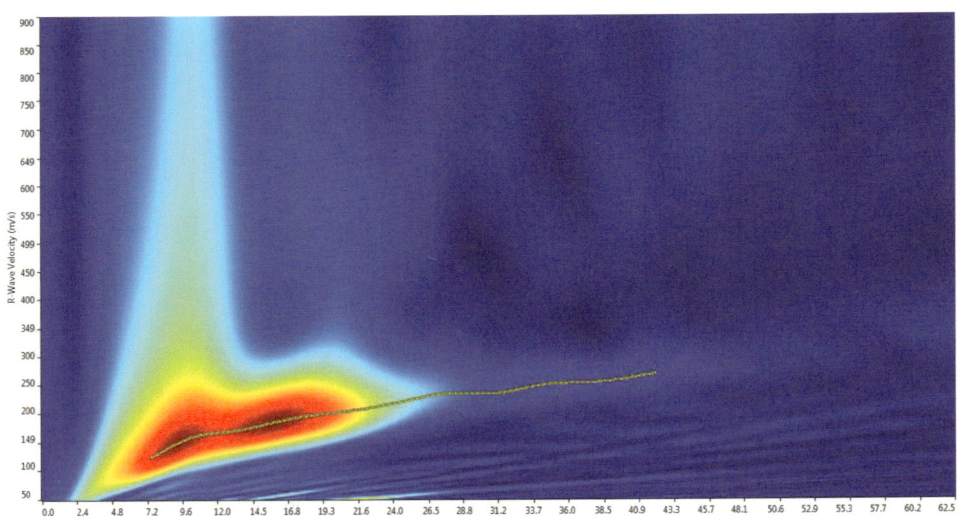

Figure 17. Picking the fundamental mode of surface layer.

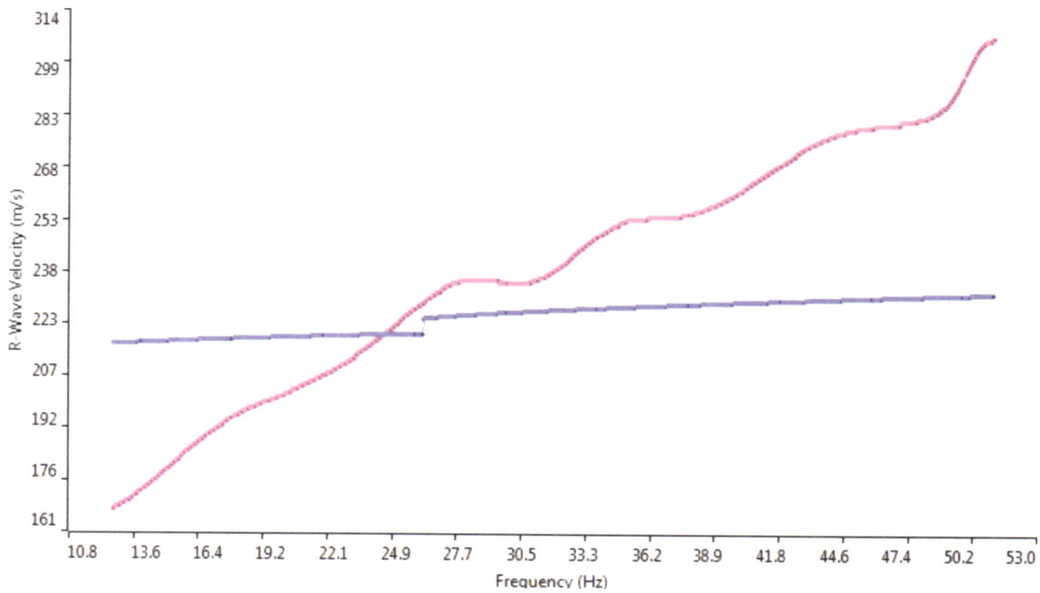

Figure 18. The dispersion curve of surface layer matching.

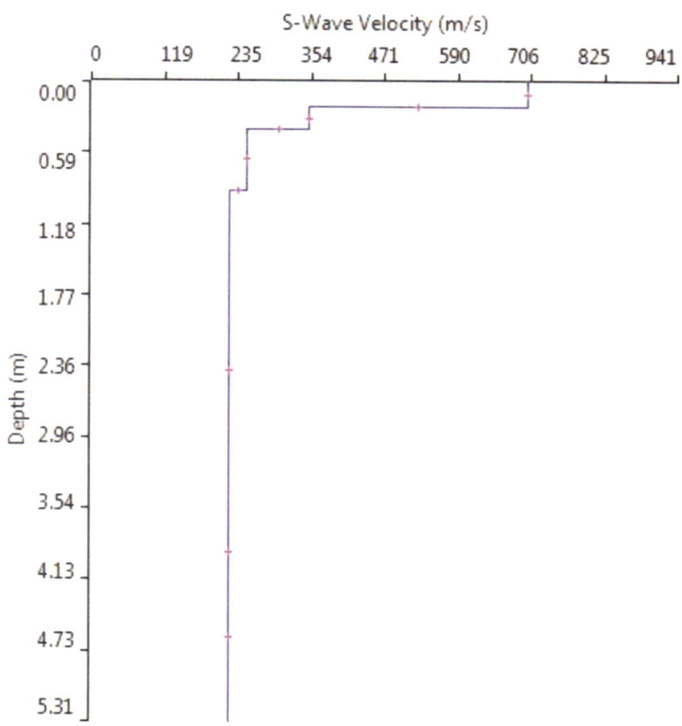

Figure 19. The profile of surface layer of the pavement.

The second branch of spectrum corresponding to the fundamental mode of surface waves from deeper layer was picked as follows (**Figures 20–23**):

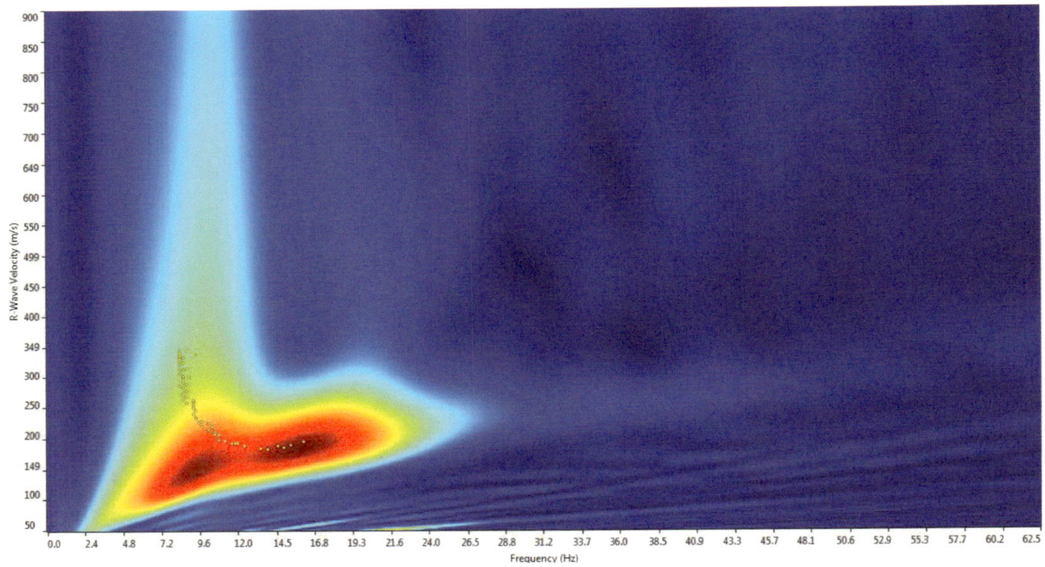

Figure 20. Picking the fundamental mode of surface waves from deeper layer.

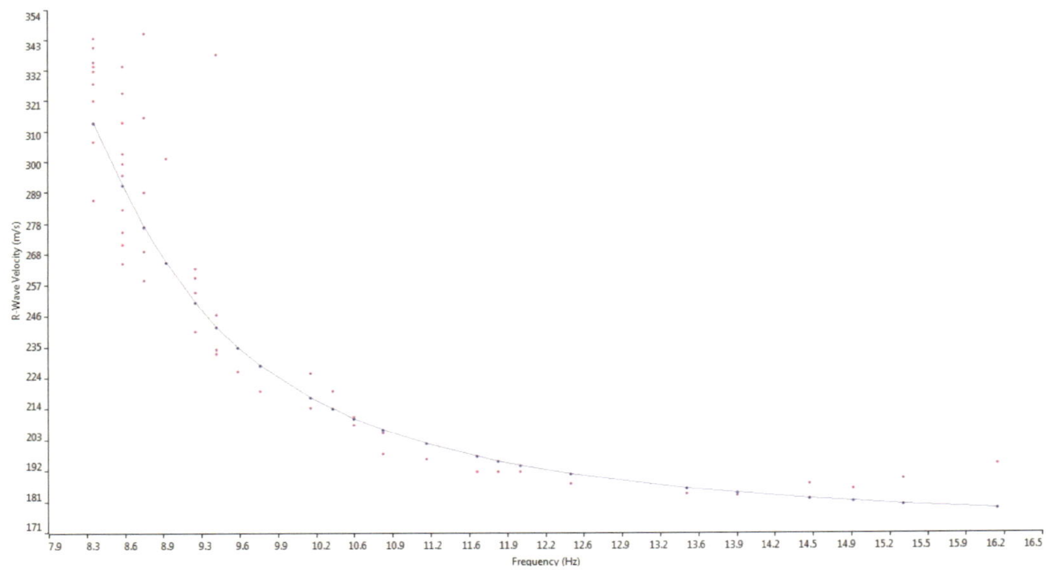

Figure 21. Curve matching of dispersion curve from deeper layer.

Tables 1 and **2** were combined to come up with **Table 3**, which is the complete shear wave velocity profile of the pavement.

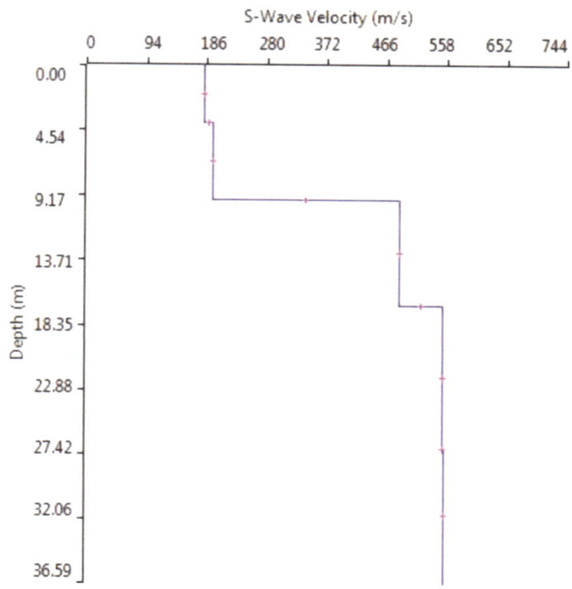

Figure 22. Profile of shear wave velocity with respect to depth of deeper layer (subgrade).

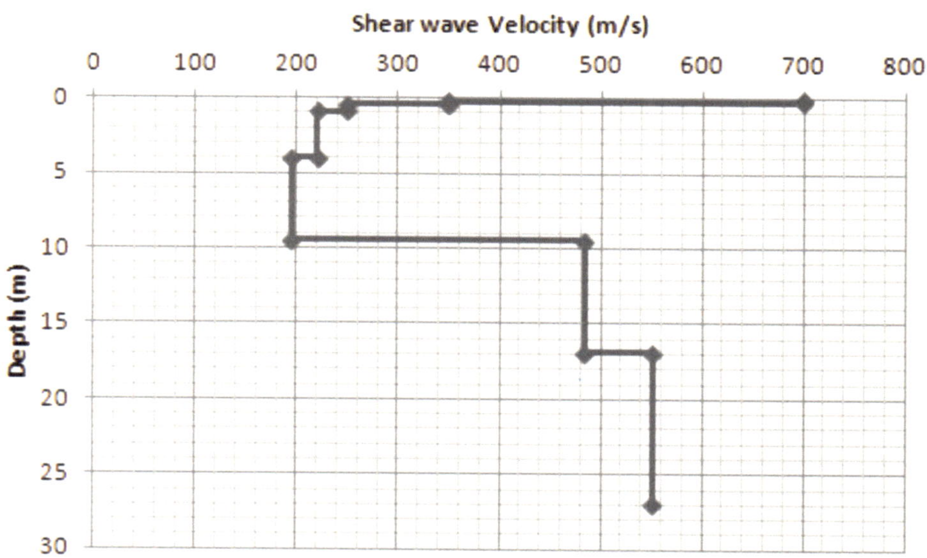

Figure 23. Profile of complete shear wave velocity of the pavement.

Thickness(m)	Depth (m)	Vs (m/s)	Vp (m/s)	Poisson	Density
0.2	0.0	700.0	1399.0	0.33	1.8
0.2	0.2	350.0	699.0	0.33	1.8
0.5	0.4	250.0	500.0	0.33	1.8
3.0	0.9	220.0	440.0	0.33	1.8
	3.9	220.0	440.0	0.33	1.8

Table 1. Results of inversion process of shallow layers.

Thickness (m)	Depth (m)	Vs (m/s)	Vp (m/s)	Poisson	Density
4.0	0.0	182.0	364.0	0.33	1.8
5.5	4.0	195.0	390.0	0.33	1.8
7.4	9.5	484.0	967.0	0.33	1.8
10.1	16.9	551.0	1101.0	0.33	1.8
	27.0	554.0	1107.0	0.33	1.8

Table 2. Results of inversion process of deeper layers.

Thickness (m)	Depth (m)	Vs (m/s)	Vp (m/s)	Poisson	Density (gr/cc)
0.2	0	700	1399	0.33	1.8
0.2	0.2	350	699	0.33	1.8
0.5	0.4	250	500	0.33	1.8
3	0.9	220	440	0.33	1.8
5.5	4	195	390	0.33	1.8
7.4	9.5	484	967	0.33	1.8
10.1	16.9	551	1101	0.33	1.8
	27	554	1107	0.33	1.8

Table 3. Results of combination of processes of shallow and deeper layers.

3. Conclusion

The MASW (Multichannel Analysis of Surface Waves) as nondestructive test for highway engineering was described. A real application to road on North Jakarta was discussed. The measurement was carried out on busy street, but the measurement was success. The results in terms of elastic modulus and thickness of each pavement layer could be used for calculating pavement service life using finite element analysis. The disadvantage of this method is that the thickness and shear wave velocity of the layers is along long line, not one point thickness and velocity. In some application such as in construction inspection, it is desired to have the thickness and velocity at one point.

Author details

Gunawan Handayani

Address all correspondence to: gunawanhandayani@gmail.com

Bandung Institute of Technology, Bandung, Indonesia

References

[1] Miller RD, Xia J, Park CB, Ivanov JM. Multichannel analysis of surface waves to map bedrock. The Leading Edge. 1999;**18**:1392-1396

[2] Park CB, Ivanov J, Miller RD, Xia J, Nils Ryden. Seismic investigation of pavements by MASW method – Geophone approach. SEG Library. 2001

[3] Multichannel Analyses of Surface Waves (MASW). What is dispersion? Available from: http://www.masw.com/Whatisdispersion.html [Accessed: June 15, 2017]

[4] *WaveEq*, version 2.3.0.5, tutorial. Available from: Text_surface_wave_method.pdf

[5] Multichannel Analyses of Surface Waves (MASW). Dispersion Imaging Scheme. n.d. Available from: http://www.masw.com/DCImagingScheme.html [Accessed: June 15, 2017]

[6] Available from: http://www.kgs.ku.edu/software/*surfseis*/masw.html [Accessed: September 7, 2017]

[7] Park CB, Miller RD. Passive multichannel analysis of surface waves (MASW). Journal of Environmental and Engineering Geophysics. 2008;**13**:1-11

[8] "*Swan*", version 1.4, Geostudi Astier via S.r.l. via A Nicolodi, 48

[9] Multichannel Analyses of Surface Waves (MASW). Data Acquisition. n.d. Available from: http://www.masw.com/DataAcquisition.html [Accessed: June 15, 2017]